An Introductory Guide to Computational Methods
for the Solution of Physics Problems

George Rawitscher (1928–2018)
Victo dos Santos Filho · Thiago Carvalho Peixoto

An Introductory Guide to Computational Methods for the Solution of Physics Problems

With Emphasis on Spectral Methods

 Springer

George Rawitscher (1928–2018)
University of Connecticut
Storrs, CT, USA

Victo dos Santos Filho
H4D Scientific Research Laboratory
Bela Vista, São Paulo, Brazil

Thiago Carvalho Peixoto
Federal Institute of Sergipe (IFS)
Nossa Senhora da Glória, Sergipe, Brazil

ISBN 978-3-319-42702-7   ISBN 978-3-319-42703-4   (eBook)
https://doi.org/10.1007/978-3-319-42703-4

Library of Congress Control Number: 2018950796

This Springer imprint is published by the registered company Springer Nature Switzerland AG
The registered company address is: Gewerbestrasse 11, 6330 Cham, Switzerland

Fig. 1 The joy of teaching   Fig. 2 The joy of learning

This book is dedicated to all the persons who love to learn and teach. The pictures show G. Rawitscher's grandson Uly and his friend Olive as preschoolers. Here they are teaching and learning, and enjoying being together.

Preface

The main purpose of this monograph is to provide an introduction to several numerical computational methods for solving physics problems. At the same time, it serves as an introduction to more advanced books, especially Trefethen [1] and Shizgal [2], which we have frequently referenced. Both of these books use MATLAB in their numerical examples, which is also the case for our present work. This monograph may then be seen as a good text for a course titled, for example, "Introduction to Numerical Methods for Undergraduate Quantum Mechanics".

Our text provides many examples of computational methods applied to solving physics problems. These include expansions into a set of Sturmian functions (Chap. 11), the iterative calculation of eigenvalues for a particular differential equation (Chap. 10), the phase–amplitude description of a wave function (Chaps. 8 and 12), the solution of a third-order differential equation (Chap. 12), the finite element method to solve a differential equation (Chap. 7), the transformation of a second-order differential equation into an integral equation (Chap. 6), or the use of expansions into Chebyshev polynomials (Chaps. 5 and 6), while keeping an eye on accuracy and convergence properties (Chap. 4). Chapters 3–6 describe "spectral" expansions, and are included because such expansions provide very accurate results and are not usually taught in courses on computational methods.

The present monograph gathers together these various computational methods that are best suited for solving physics problems. We examine the errors of such methods with many examples and show that spectral methods can be faster and at the same time more accurate than finite difference methods. This is also demonstrated by means of the numerical examples in Chaps. 6 and 7. The monograph is not intended to be mathematically rigorous since so many excellent mathematically rigorous books already exist describing spectral methods [3–16]. Such methods have been used to solve many different equations, as is the case of Vlasov equation [17], Navier–Stokes equation [18], Fisher equation [19], Schrödinger and Fokker–Planck equations [20] amongst others.

Spectral methods were first introduced in the 1970s. They are more advantageous than other methods because they tend to converge quickly and generally provide high accuracy, as described in Sect. 3.4.2. However, finite difference methods (based on the Taylor series) still are of importance for specific applications [21–22], some of which we describe in Chap. 2. Spectral methods lead to matrix equations that are more complicated than those for the finite difference methods, but the spectral convergence and accuracy gained [2] easily outweigh any drawback.

In more detail, the content of this monograph is as follows: Chaps. 1 and 2 give a review of computational errors and finite difference methods, respectively. Chapters 3 and 4 describe the collocation and Galerkin methods. An advantage of these methods is demonstrated in Chap. 4 by giving theorems on the convergence of spectral expansions. Advantages are also shown in Chaps. 5 and 6, where practical examples are given for the convergence of expansions in Chebyshev polynomials. Chapter 6 describes the Lippmann–Schwinger integral equation whose solution gives better accuracy than the equivalent Schrödinger equation, and is solved with the aid of Green's functions, all in coordinate space. In Chap. 7, we compare various finite element spectral methods. Chapters 8–12 are dedicated to various computational method examples. Chapter 8 describes the phase–amplitude method and its application to physical problems involving interesting potentials. In Chap. 9, we describe the solution for the problem of finding eigenvalues iteratively in a simple example of a vibrating inhomogeneous string. Chapter 10 develops an iterative method to obtain the energy eigenvalues of the second-order differential equation for a vibrating string. A review of expansions in Sturmian functions is presented in Chap. 11. Chapter 12 provides a novel method to solve a third-order differential equation, based on spectral expansions and the implementation of the asymptotic boundary conditions without the use of Green's functions. Finally, in Chap. 13, we present our final considerations and general conclusions of the present work.

In summary, the purpose of this monograph is to provide a compact and simple introduction to several computational methods generally not taught in traditional courses, and to examine the errors and the advantages of such methods as shown in many examples. We hope that this monograph provides students and teachers with a comprehensive foundation for a smooth transition to more advanced books.

Storrs, CT, USA   George Rawitscher
São Paulo, Brazil   Victo dos Santos Filho
São Paulo, Brazil   Thiago Carvalho Peixoto
April 2018

References

1. L.N. Trefethen, *Spectral Methods in MATLAB* (SIAM, Philadelphia, 2000)
2. B.D. Shizgal, *Spectral Methods in Chemistry and Physics. Applications to Kinetic Theory and Quantum Mechanics* (Springer, Dordrecht, 2015)
3. D. Gottlieb, S.A. Orszag, *Numerical Analysis of Spectral Methods* (SIAM, Philadelphia, 1977)

4. B. Fornberg, *A Practical Guide to Pseudospectral Methods* (Cambridge Monographs on Applied and Computational Mathematics, Cambridge University Press, Cambridge, UK, 1998)
5. D. Gottlieb, J.S. Hesthaven, J. Comput. Appl. Math. **128**(1–2), 83–131 (2001)
6. J.S. Hesthaven, S. Gottlieb, D. Gottlieb, *Spectral Methods for Time-Dependent Problems* (Cambridge University Press, Cambridge, 2007)
7. J.P. Boyd, *Chebyshev and Fourier Spectral Methods* (Dover, New York, 2001)
8. C. Canuto, M. Y. Hussaini, A. Quarteroni, T.A. Zang, *Spectral Methods: Fundamentals in Single Domains* (Springer, New York, 2006)
9. M.O. Deville, P.F. Fisher, E.H. Mund, *High Order Methods for Incompressible Fluid Flow* (Cambridge University Press, Cambridge, 2002)
10. C.-I. Gheorghiu, *Spectral Methods for Differential Problems*, Casa Cartii de Stiinta, Cluj-Napoca, Romania, 2007. Available at http://www.ictp.acad.ro/gheorghiu/spectral.pdf
11. R. Peyret, *Spectral Methods for Incompressible Viscous Flow* (Springer, New York, 2002)
12. G. Ben-Yu, *Spectral Methods and Their Applications* (World Scientific, Singapore, 1998)
13. D. Funaro, *Polynomial Approximation of Differential Equations* (Springer, Berlin, 1992)
14. D.A. Kopriva, *Implementing Spectral Methods for Partial Differential Equations Algorithms for Scientists and Engineers* (Springer, Berlin, 2009)
15. C. Shu, *Differential Quadrature and Its Application in Engineering* (Springer, Berlin, 2000)
16. J. Shen, T. Tang, L.-L. Wang, *Spectral Methods: Algoritms, Analysis and Applications* (Springer, Berlin, 2011)
17. L. Gibelli, B.D. Shizgal, Spectral convergence of the Hermite basis function solution of the Vlasov equation: The free-streaming term. J. Comput. Phys. **219**(2), 477–488 (2006)
18. P.R. Spalart, R.D. Moser, M.M. Rogers, Spectral methods for the Navier-Stokes equations with one infinite and two periodic directions. J. Comput. Phys. **96**(2), 297–324 (1991)
19. D. Olmos, B.D. Shizgal, A spectral method of solution of Fisher's equation. J. Comput. Appl. Math. **193**(1), 219–242 (2006)
20. J. Lo, B.D. Shizgal, Spectral convergence of the quadrature discretization method in the solution of the Schrödinger and Fokker-Planck equations: comparison with sinc methods. J. Chem. Phys. **125**(19), 8051 (2006)
21. Randall J. LeVeque, *Finite Difference Methods for Ordinary and Partial Differential Equations, Steady State and Time Dependent Problems* (SIAM, Philadelphia, 2007)
22. L.N. Trefethen, *Finite Difference and Spectral Methods for Ordinary and Partial Differential Equations* (Cornell University, Department of Computer Science, Center for Applied Mathematics, Ithaca, NY, 1996)

Acknowledgements

The authors thank all professionals and friends who have analysed the content of the book and helped to elaborate and improve it. One of the authors (G. R.) is grateful to the International Centre for Theoretical Physics-South American Institute for Fundamental Research (ICTP-SAIFR), in particular to its director Nathan Jacob Berkovits and the Secretariat of the institute, for the invitation to teach a mini-course on spectral computational methods in São Paulo, Brazil. The course took place from 16 March to 26 April 2015 and consisted of twelve lectures. These lectures provided the initial stimulus for writing this monograph, in cooperation with the two Brazilian co-authors (V. S. F. and T. C. P.) who took the course and contributed significantly to writing the text. G. R. is very grateful to Profs. Lauro Tomio and Sadhan K. Adhikari for enthusiastically supporting the visit and for their dedicated hospitality. G. R. is also much indebted to his parents and his wife Joyce, who inspired in him throughout his life a spirit of freedom and accomplishment. The author V. S. F. thanks God and his parents for always supporting and helping him in his life. V. S. F. also thanks each member of his family and all of his friends (with special acknowledgement to Prof. Lauro Tomio) who encouraged or helped him in the process of writing this monograph.

ICTP-SAIFR is a South American Regional Centre for Theoretical Physics created through a collaboration of the Abdus Salam International Centre for Theoretical Physics (ICTP) with the São Paulo State University (UNESP) and the São Paulo Research Funding Agency (FAPESP). The author G. R. would like to recognize FAPESP grant 2011/11973-4 for funding his visit to ICTP-SAIFR.

Finally, the authors thank Peter and Henry Rawitscher for their valuable help of seeing through the publication and in proofreading the text and correcting the english grammar, which greatly improved our monograph.

Contents

In Memoriam

In Memoriam by Peter and Henry Rawitscher

Professor George Rawitscher, our father, passed away on 10 March 2018. He was a refugee from Germany in the 1930s who moved to Brazil as a small child. He attended the University of São Paulo, where he had the opportunity to study under Richard Feynman. George Rawitscher then did a Ph.D. at Stanford, and was a postdoc at Yale University later. He spent the bulk of his career at the University of Connecticut.

George Rawitscher developed a lifelong bond with physics at a young age. His career at the University of Connecticut involved both his own research and teaching. He equally was happy to teach the basic classes as well as advanced, as he enjoyed sharing knowledge with all people. He continued the path of physics for over 70 years, still writing papers and doing research into his ninth decade. His mind was clear and full of knowledge until his last days. He was also known to be compassionate to most. He enjoyed human interaction and would always respond. He liked to visit Brazil occasionally.

One goal that George Rawitscher had in his last days was the completion of this book.

The family thanks Prof. Victo dos Santos Filho for bringing that goal, this book, to fruition with his hard work.

In Memoriam by Victo dos Santos Filho

My opportunity to work with Prof. George Rawitscher arose when he was giving a lecture series at the Institute of Theoretical Physics (IFT) of Sao Paulo State University (UNESP). His host was our mutual friend Prof. Lauro Tomio, who assisted Prof. Rawitscher and his wife during that time. Professor Tomio told me about the lectures and, as I had a personal interest in computational physics and

advanced numerical methods applied to physics problems, I decided to attend his course. The picture shows Prof. Rawitscher at that time.

Professor Rawitscher's classes impressed me with their depth of content and quality. At the end of the lecture series, he expressed his desire to publish his teachings in a book. I appreciated his strong energy and wisdom and realized that I could learn much by working with him. I myself and graduate student Thiago Carvalho Peixoto accepted to collaborate on the book project with him. This challenge proved to be a great experience, giving us the opportunity to work together with an impressively dedicated physicist.

Now that Prof. Rawitscher has passed away before his dream of seeing the book published came true, I would like to pay tribute not only to the scientist but to the person who George Rawitscher was. In addition to being a serious professional and a great physicist, with vast experience and knowledge in his area, he was also a great man. I was physically present with him for a short time and then worked with him at a distance for a longer period. During that time, I shared good moments and conversations and I could see in him some characteristics that are very rare in many human beings: he possessed kindness, serenity, wisdom, equilibrium and gentleness. These qualities were evident in his simpler attitudes and words.

I have a philosophy of life that has been my guiding star: whenever possible, live with people and work with professionals who have the quality of being serious, wise, honest and good hearted, all of which describe very well Prof. Rawitscher. So, I thank God for the opportunity of having worked with him on the present monograph.

To George Rawitscher, my wish is that he is at peace and happy with his wife in their new life in Heaven.

Chapter 1
Numerical Errors

Abstract In this chapter, we illustrate the occurrence of round-off errors when specific recurrence relations are used. Round-off errors can occur in numerous other types of computations such as in the calculation of Bessel or Coulomb functions, and we will try to convince the reader that such errors are unavoidable in any numerical calculation.

1.1 The Objective and Motivation

Each numerical method contains two types of errors: the "truncation error" and the "round-off errors". The truncation error originates from the approximations to an analytical solution, and the "round-off errors" arise from the discreteness and finiteness of the numbers which the computer can carry. An excellent discussion of round-off error in numerical computation is provided in Chapter 1 of the book "Scientific Computing with MATLAB" by Quarteroni, Sleria and Gervasio [1] as well as in many other textbooks. The most common truncation errors occur when making an expansion in an infinite number of terms, but necessarily truncating the expansion into a finite number of terms. The round-off errors accumulate during the calculation, and could become overwhelmingly large if the computation involves too many numbers of steps. Finding a good balance between these two sources of errors is the challenge that makes computational methods interesting.

The size of the round-off errors depends on the number of significant figures employed by the computer (8 or 9 for simple precision Fortran, 15 for double precision, 16 for MATLAB, etc.) and hence the digits not available to the computer accumulate as errors, called "round-off errors". Another good discussion of round-off errors is given in Section 1.2 of Ref. [2]. For spectral methods the truncation errors are due to the fact that expansions of the solution to a particular equation into any particular set of basis functions have to be truncated at some upper value when carried out numerically by computer. These left out terms represent the truncation error. The larger they are, the corresponding truncation error is similarly large. For the finite difference methods the truncation errors occur when a Taylor series expansion of a function, which has an infinite number of terms, is truncated at a finite

© Springer Nature Switzerland AG 2018

G. Rawitscher et al., *An Introductory Guide to Computational Methods for the Solution of Physics Problems*,

https://doi.org/10.1007/978-3-319-42703-4_1

number of terms. Numerical errors also occur with recursion relations, iterations, or successive approximations, as shown in next section, whether they converge or not. In subsequent chapters expansions of various functions into a set of particular basis functions are developed so as to verify the theorems on the rapidity of convergence of such expansions [3]. The present chapter offers some numerical examples of accuracy losses.

1.2   Accuracy in Numerical Calculations

Besides the occurrence of errors in numerical calculations, there also are two additional issues in any numerical approach: namely, what is meant by "stability" and "accuracy".

An instability can occur regardless of the number of significant figures employed in the calculation, and depends on the nature of the equation to be solved. The accuracy of a calculation is measured in terms of the error of the numerical result, compared to a known exact result. The instability manifests itself through the deterioration of the accuracy as the calculation proceeds. A good analysis of the concept of stability, as it applies to various algorithms, is extensively discussed in Ref. [2].

For example, in Project 1.1 the recurrence relation Eq. (1.1) has two theoretical solutions, one which decreases as the number of recursion cycles increases and the other which increases. The second solution will numerically mix itself into the first solution, and will make the whole recurrence process unstable if carried beyond a certain number of recurrence cycles.

Another example is presented in Project 1.2. The recursion relations given by Eqs. (1.6) and (1.7) are not only satisfied by the function $J_v(x)$, but also by the function $Y_v(x)$, $v = 0, 1, 2, \ldots$, as defined in Ref. [4], Eq. (9.1.27), or in Ref. [3], Eq. (3.6.11). The function J is small at small values of x, while the function Y is large, as illustrated in Fig. 9.2 in Ref. [4]. Hence, due to numerical round-off errors, the function Y will manifest itself numerically and introduce an error in the calculation of J, depending on whether the recursion is carried forward or backward. Such types of instabilities also occur in the so called "stiff" equations, as discussed for example in Chapter 8.14 of Ref. [5]. The effect of truncation errors will be illustrated in Chap. 2, in connection with the solution of first or second order differential equations by means of finite difference methods, while in the examples 1.1 and 1.2 there are no truncation errors.

1.2.1   Assignments

For all assignments please write a short essay explaining what your results show and what you have learned in the process.

Fig. 1.1 A solution to
Eq. (1.1) of Project 1.2,
calculated via a MATLAB
program

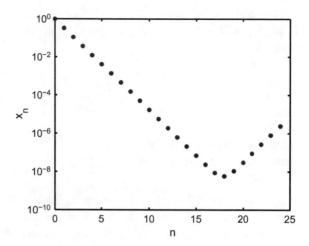

Project 1.1: Recursion Relation

Assignment 1.1: Consider the recursion relation

$$x_n = \frac{10}{3}x_{n-1} - x_{n-2}, \qquad (1.1)$$

with $x_0 = 1$, and $x_1 = 1/3$.

Examine numerically and theoretically whether this recursion is stable or unstable.

Assignment 1.2: Recursion Relation

Repeat Assignment 1.1, but using a different recursion relation

$$x_n = \frac{5}{2}x_{n-1} - x_{n-2}, \qquad (1.2)$$

with $x_0 = 1$ and $x_1 = 1/2$.

If programmed with MATLAB (with 16 significant figures) the result for Assignment 1.1 should look like Fig. 1.1. Assuming that you found a similar result, please explain why the value of x started to increase again for $n > 17$. Please note that for this example the "sweet spot" is $n \simeq 17$. The sweetspot is the place where the numerical truncation errors are approximately the same as accumulation of round-off errors. At this point the errors can not be further reduced.

Hint: This question can be analyzed by considering the general two step recursion relation

$$x_n = a\, x_{n-1} + b\, x_{n-2} \qquad (1.3)$$

(where n is the iteration index) for which there are two independent solutions y_n and z_n:

$$x_n = \alpha y_n + \beta z_n. \qquad (1.4)$$

Assuming that $y_n = \lambda_1 y_{n-1} = \lambda_1^2 y_{n-2}$ and similarly $z_n = \lambda_2 z_{n-1} = \lambda_2^2 z_{n-2}$, then x_n will satisfy Eq. (1.4) provided that each λ satisfies

$$\lambda^2 = a\lambda + b. \tag{1.5}$$

For the example $a = 10/3$ and $b = -1$, the corresponding quadratic equation $\lambda^2 = (10/3)\lambda - 1$ has the two roots $\lambda_1 = 1/3$ and $\lambda_2 = 3$. For the exact conditions $x_0 = 1$, and $x_1 = 1/3$ one obtains $\alpha = 1$, and $\beta = 0$, and hence, according to Eq. (1.4), $x_2 = (1/3)^2$, $x_3 = (1/3)^3, \ldots$. This hint should be a sufficient guide for the explanation of the behavior shown in Fig. 1.1, and for your result of Assignment 1.2.

Project 1.3: Bessel Function Recursion Relations
Consider the Bessel Functions $J_\nu(x)$, where $\nu = 0, 1, 2, \ldots$ is the index, and x is the variable. These functions obey the upward recursion relations

$$J_{\nu+1}(x) = -J_{\nu-1}(x) + 2(\nu/x)J_\nu(x); \quad \nu = 1, 2, 3, \ldots, \tag{1.6}$$

and the downward recursion relations

$$J_{\nu-1}(x) = -J_{\nu+1}(x) + 2(\nu/x)J_\nu(x); \quad \nu = \nu \max, \nu \max -1, \nu \max -2, \ldots, 1. \tag{1.7}$$

Explore numerically whether these two recursion relations are stable for $x_0 = 0.2$. For Eq. (1.6) use for J_0 and J_1 the numerical values given in the Table 1.1, or else that obtained by some other analytic expression, and for Eq. (1.7) choose a value for ν_{max}, for example $\nu_{max} = 10$, and use the value given in the Table 1.1. For larger values of ν_{max} use MATLAB to calculate $J_{\nu \max}$ and $J_{\nu \max -1}$. The "stability" is measured in terms of the rapidity of accumulation of the relative errors. The errors are obtained by comparing the result of the recursion relation with the values of the Bessel functions that are available in the literature. An extensive discussion of recurrence relations is given in section 2.2 of Ref. [6], with extensive tables indicating the respective errors.

Table 1.1 Numerical values of Bessel function

ν	$J_\nu(x = 0.2)$
0	0.990024972239576
1	0.099500832639236
2	0.00498335415278357
3	0.000166250416435268
4	4.15834027447194e-006
5	8.31945436094694e-008
6	1.38690600152496e-009
7	1.98164820280361e-011
8	2.47740437568484e-013
9	2.75297744273373e-015
10	2.7532277551303e-017

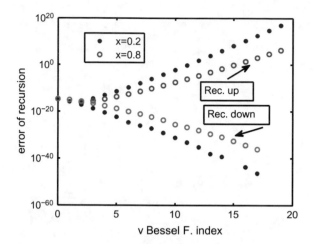

Fig. 1.2 Error of the upward and downward recursion relations for Bessel functions $J_\nu(x)$. The upward recursions, illustrated by the upper two sets of points, start from the known functions $J_\nu(x)$ with $\nu = 0$ and $\nu = 1$ (given in the list of functions in the text). The downward ones, illustrated by the lower two sets of points, start with $\nu = 19$ and $\nu = 18$, obtained from MATLAB. The closed symbols correspond to $x = 0.2$, while the open ones correspond to $x = 0.8$. In the present example these starting values are obtained from MATLAB. Obviously the recursion "down" is stable

Note that in MATLAB [7] some specialized mathematical functions, such as Bessel functions, Legendre polynomials, or gamma functions, can be found by calling "help specfun". Bessel functions can be called by $bessel j\, (\nu, x)$, where ν can also be non-integer, and x can be a vector. For instance, when $x = 0.2$, we have the numerical values shown in Table 1.1. Your result should look like Fig. 1.2.

References

1. A. Quarteroni, F. Saleri, P. Gervasio, *Scientific Computing with MATLAB and Octave* (Springer, Berlin, 2014). ISBN 978-3-642-45366-3
2. W.H. Press, S.A. Teukolsky, W.T. Vetterling, B.P. Flannery, *Numerical Recipes: The Art of Scientific Computing (Fortran Version)* (Cambridge University Press, New York, 1990). ISBN 0-521-38330 7
3. W.J. Olver, D.W. Lozier, R.F. Boisvert, C.W. Clark, *NIST Handbook of Mathematical Functions*, National Institute of Standards and Technology, U.S. Department of Commerce (Cambridge University Press, Cambridge, 2010)
4. M. Abramowitz, I. Stegun (eds.), *Handbook of Mathematical Functions* (Dover, New York, 1972)
5. A. Gilat, V. Subramaniam, *Numerical Methods for Engineers and Scientists*, 2nd edn. (Wiley, New York, 2011)
6. B.D. Shizgal, *Spectral Methods in Chemistry and Physics. Applications to Kinetic Theory and Quantum Mechanics* (Springer, Dordrecht, 2015)
7. Learning MATLAB 6.5 (by MathWorks, 2002). ISBN-10: 0967219574

Chapter 2
Finite Difference Methods

Abstract In this chapter, we describe two numerical finite difference methods which are used for solving differential equations, e.g., the Euler method and Euler-Cromer method. The emphasis here is on algorithm errors, and an explanation of what is meant by the "order" of the error. We show that the Euler method introduces an error of order 2, denoted as $\mathscr{O}(2)$, while the latter presents errors of order $\mathscr{O}(3)$. We finish the chapter by applying the methods to two important physical problems: the physics of the pendulum and the physics of descending parachutes.

2.1 The Objective and Motivation

A brief review of the finite difference method in its simplest form will be presented here, in order to provide a contrast to the spectral methods described in Chaps. 3–5. These finite difference methods are based on Taylor's expansion of the solution, and are denoted in Ref. [1], section 3.7, as Taylor Series Methods. First the finite difference method in general will be reviewed, and two numerical applications are provided subsequently. Finite difference methods for solving a differential or integral equation were introduced in the 1950s. They are taught in most elementary numerical methods courses, and they have many applications [2, 3]. They can also be applied to solving the Schrödinger equation [4] numerically. This is a wave equation which describes the quantum nature of a particle, and has been in existence since 1926 [5], and we still are searching for better methods to solve it.

2.2 Order of the Methods

Finite difference methods [6–8] are based on the Taylor expansion of a function $f(x)$

$$f(x + h) = f(x) + hf'(x) + \frac{h^2}{2}f''(x) + \frac{h^3}{6}f'''(x) + \cdots \frac{h^n}{n!}f^{(n)}(x) + \cdots, \tag{2.1}$$

© Springer Nature Switzerland AG 2018
G. Rawitscher et al., *An Introductory Guide to Computational Methods for the Solution of Physics Problems*,
https://doi.org/10.1007/978-3-319-42703-4_2

where the primes denote differentiation with respect to x. This expansion is described for example in Ref. [1], Eq. (1.4(vi)). For an equidistant set of mesh points separated by the distance h, a convenient notation is $x_1 = x$, $x_2 = x + h, \ldots, x_n = x + (n-1)h, \ldots$, and the above equation takes the form

$$f_{n+1} = f_n + h f_n' + \cdots \frac{h^n}{n!} f_n^{(n)} + \mathcal{O}(h^{n+1}), \ n = \pm(1, 2, \ldots). \tag{2.2}$$

In Eq. (2.2), $f_n = f(x_n)$, and $\mathcal{O}(h^{n+1})$ denotes the remainder of the expansion that was truncated at order n. Based on the above expressions one can show that

$$f_n' = \frac{f_{n+1} - f_n}{h} + \mathcal{O}(h), \tag{2.3}$$

while a smaller error for the derivative, of order h^2, is given by

$$f_n' = \frac{f_{n+1} - f_{n-1}}{2h} + \mathcal{O}(h^2). \tag{2.4}$$

Similar equations are also given by [9] in section 3.9.1.

The simplest procedure to solve a second order differential equation numerically is to use the Euler method, which will be explained by means of an example given next. The accuracy and stability of this method is investigated extensively in Ref. [8], Chapter 8. An application to Newton's equation of motion will now be described [10]. These equations describe the motion of a particle of mass m under the influence of a force $\mathbf{F}$, according to which the acceleration equals the force divided by the mass. In this case the variables in the Taylor expansions above are modified as follows: x is replaced by the time t, the time increment is τ (taking the place of h), the position vector is $\mathbf{r}$, the velocity is $\mathbf{v}$, the acceleration is $\mathbf{a}$ and one has

$$\frac{d\mathbf{v}}{dt} = \mathbf{a} \quad \text{and} \quad \frac{d\mathbf{r}}{dt} = \mathbf{v}. \tag{2.5}$$

The quantities, designated by bold letters, have three dimensions in x, y, and z. The fact that the acceleration can be a function of both position and velocity can be expressed as $\mathbf{a}[\mathbf{r}, \mathbf{v}]$. Since $\mathbf{a} = d\mathbf{v}/dt$, and making use of Eq. (2.3) one obtains the velocity and position at a future time step $t + \tau$ according to

$$\mathbf{v}(t + \tau) = \mathbf{v}(t) + \tau \, \mathbf{a}[\mathbf{r}(t), \mathbf{v}(t)] + O(\tau^2), \tag{2.6}$$

$$\mathbf{r}(t + \tau) = \mathbf{r}(t) + \tau \, \mathbf{v}(t) + O(\tau^2). \tag{2.7}$$

In Eqs. (2.6) and (2.7) it is assumed that the values of $\mathbf{r}$, $\mathbf{v}$, and $\mathbf{a}$ are already known at time t from the previous steps, and we are now looking for the values at $t + \tau$ for the next step. These equations in the time discretized form can be written as

$$\mathbf{v}_{n+1} = \mathbf{v}_n + \tau \, \mathbf{a}_n + O(\tau^2), \tag{2.8}$$

$$\mathbf{r}_{n+1} = \mathbf{r}_n + \tau \, \mathbf{v}_n + O(\tau^2). \tag{2.9}$$

The initial conditions at $t = 0$ are then given by the values of $\mathbf{v}_1 = \mathbf{v}(t = 0)$ and $\mathbf{r}_1 = \mathbf{r}(t = 0)$, and by successive implementation of Eqs. (2.8) and (2.9), the values of $\mathbf{v}$ and then subsequently $\mathbf{r}$ at times t_n, $n = 2, 3, \ldots$, can be obtained. The main question to be answered is what size does τ have to be so as to achieve a given accuracy in a given time interval. Equations (2.8) and (2.9) represent the method.

The Euler-Cromer method [7] consists in keeping Eq. (2.8), but replacing Eq. (2.9) by

$$\mathbf{r}_{n+1} = \mathbf{r}_n + \tau \, \mathbf{v}_{n+1} + O(\tau^2). \tag{2.10}$$

An improvement of Eq. (2.10) consists in replacing $\mathbf{v}_{n+1}$ in that equation by $(1/2)(\mathbf{v}_n + \mathbf{v}_{n+1})$, which is equivalent to (can you show this?)

$$\mathbf{r}_{n+1} = \mathbf{r}_n + \tau \, \mathbf{v}_n + \frac{1}{2}\mathbf{a}_n \tau^2 + O(\tau^3). \tag{2.11}$$

There is also the so called "leap-frog" method. It achieves a smaller truncation error by using Eq. (2.4)

$$\frac{\mathbf{v}_{n+1} - \mathbf{v}_{n-1}}{2\tau} = \mathbf{a}(r_n) + O(\tau^2)$$

and

$$\frac{\mathbf{r}_{n+2} - \mathbf{r}_n}{2\tau} = \mathbf{v}_{n+1} + O(\tau^2),$$

which can be written as

$$\mathbf{v}_{n+1} = \mathbf{v}_{n-1} + 2\tau \, \mathbf{a}_n + O(\tau^3) \tag{2.12}$$

and

$$\mathbf{r}_{n+2} = \mathbf{r}_n + 2\tau \, \mathbf{v}_{n+1} + O(\tau^3). \tag{2.13}$$

If the force at position n is known, then $\mathbf{a}_n$ will be known, and hence $\mathbf{v}_{n+1}$ can be calculated from Eq. (2.12). Subsequently $\mathbf{r}_{n+2}$ can be obtained from Eq. (2.13) since $\mathbf{v}_{n+1}$ is now known. An application to the motion of a pendulum is described in next section. There, the relation between position and acceleration is non linear, and hence an "implicit" method has to be used [8].

Other more sophisticated methods are commonly used. Among them is the Runge–Kutta method, described in Ref. [11], Chapter 8, Section 8, which has a truncation error of $O(\tau^5)$, and the Numerov method with a truncation error of $O(\tau^6)$. The latter is also denoted as Milnes' corrector method, and is especially useful if the relation between acceleration and position is linear. These methods are described in Eqs. (25.5.22) and (25.5.21C) of Ref. [12], respectively.

2.3   The Physics of the Frictionless Pendulum

Consider a mass m attached to the lower end of a string of length ℓ, and denote by θ (in radians) the angle which the string makes with the vertical direction. When lifted to an initial angle θ_0 and then released from rest, the trajectory of the mass is an arc of a circle located in the vertical plane. By considering the component of the weight of m along the tangent of that circle (given in magnitude by $mg\ell \sin\theta$) and by remembering that tangential force equals the mass times tangential acceleration, one obtains the equation of motion of the angle θ

$$\frac{d^2}{dt^2}\theta = -\frac{g}{\ell}\sin(\theta). \tag{2.14}$$

Note that $\sqrt{\ell/g}$ has the dimension of time (g is the acceleration of gravity in m/s^2), and $\sqrt{\ell/g}$ is the natural unit of time for this problem. By going to the dimensionless variable $\bar{t} = t/\sqrt{\ell/g}$, and by denoting $f' = df/d\bar{t}$, the equation of motion (2.14) becomes identical to

$$\theta'' = -\sin(\theta), \tag{2.15}$$

where $\theta'' = d^2\theta/d\,\bar{t}^2$. We leave it to the reader to show that from conservation of energy one obtains

$$\theta' = \pm 2[\sin^2(\theta_0/2) - \sin^2(\theta/2)]^{1/2}, \tag{2.16}$$

where θ_0 is the initial angle (Hint: the height of the pendulum above its lowest height is $\ell(1 - \cos\theta)$. This angle θ_0 is also the maximum angle, since the pendulum is released from rest at this angle. Every half cycle the sign in Eq. (2.16) changes.

Using a finite difference solution of Eq. (2.14), one obtains the time dependence of θ displayed in Fig. 2.1.

In this figure the analytic solution of the linear approximation to Eq. (2.15)

$$\theta_A'' = -\theta_A \tag{2.17}$$

is also shown for comparison purposes. The solution of Eq. (2.17) is given by $\theta_A = \alpha \sin(\bar{t}) + \beta \cos(\bar{t})$, where the constants α and β are determined by the initial conditions. In the present case one finds $\alpha = 0$, $\beta = \theta_0$. The figure shows that the physical pendulum takes a longer time to reach maximum angular displacement than the analytical pendulum. This physical pendulum is also denoted a "simple" pendulum, because the effect of friction is not included. Friction leads to very interesting damping phenomena, that are discussed extensively in other books [10, 13].

In the present method of calculation energy is nearly conserved. This can be seen by observing that the maximum of θ is very close to θ_0 in Fig. 2.1. An algorithm that is more suitable to satisfy energy conservation is given by the Verlet method [14], as follows: the equation to be solved is $x'' = f(t, x)$, where a prime denotes a

Fig. 2.1 Comparison of three solutions for the pendulum. The time is dimensionless $\bar{t} = (g/\ell)^{1/2} t$. The solutions of the various orders h^2 and h^3 are based on various forms of finite difference methods. The analytical result, valid if $\sin\theta$ is replaced by θ (red solid line), is still in reasonable agreement with the numerical result (green line with "o"), for which this approximation is not made. Please note that energy conservation holds quite well

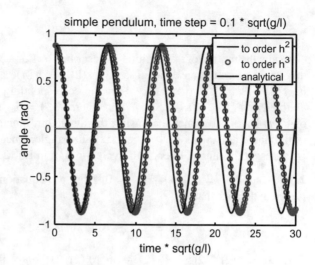

derivative with respect to time t, and the function f arises from Newton's equations. If the numerical values of x and x' at times $t_k, k = 1, 2, \ldots$, are denoted as w_k and u_k, respectively, and the spacing of the t_k values is denoted as h, then Verlet's algorithm is given by

$$w_{k+1} = w_k + h\, u_k + \frac{h^2}{2} f(t_k, w_k), \tag{2.18}$$

$$u_{k+1} = u_k + \frac{h}{2}\left[f(t_k, w_k) + f(t_{k+1}, w_{k+1})\right].$$

By comparison, Euler's method is given by

$$w_{k+1} = w_k + h\, u_k + \frac{h^2}{2} f(t_k, w_k), \tag{2.19}$$

$$u_{k+1} = u_k + h\, f(t_k, w_k).$$

Project 2.1: Explore the validity of Eqs. (2.3) and (2.4) by means of the example $f(x) = \sin(x)$ for $x \in [0, \Pi/2]$. The objective is to verify whether the errors $\mathcal{O}(h)$ and $\mathcal{O}(h^2)$ of the derivative really are proportional to h and h^2, respectively.

(a) This can be achieved by evaluating both Eqs. (2.3) and (2.4) for a fixed value of x and a range of values of h. We suggest that a plot of $\mathcal{O}(h)$ and $\mathcal{O}(h^2)$ as a function of h will be very informative. In MATLAB the command log log is helpful.

(b) What happens when h becomes very small? Please provide an explanation of your result.

2.3.1 Assignments

Consider the motion of a pendulum, as described by Eqs. (2.14)–(2.16). In this assignment the time values are denoted as $t(1), t(2), \ldots, t(n), \ldots$, where $t(1) = 0$, $t(2) = h$, $t(3) = 2h, \ldots, t(n) = (n-1)h$; the corresponding values of the angle θ are denoted as $f(1), f(2), \ldots, f(n), \ldots$, where $f(1) = \theta_0$, and where the other values of $f(n)$ have to be calculated using the algorithm indicated below. Here θ_0 is the initial angle at t_1. This is also the maximum angle, since the pendulum is released from rest at this angle. Every half cycle the sign in Eq. (2.16) changes.

2.1: Using $f(1) = \theta_0$, and $f'(1) = 0$, from the equations earlier given show with pencil and paper that

$$f(2) = f(1) - \sin(f(1))h^2/2 + \mathcal{O}(h^4), \tag{2.20}$$

$$f(3) = f(2) - 2[\sin^2(\theta_0/2) - \sin^2(f(2)/2)]^{1/2}h - \sin(f(2))h^2/2 + \mathcal{O}(h^3), \tag{2.21}$$

and in general for $n \geq 2$ show that

$$f(n+1) = f(n) \pm 2[\sin^2(\theta_0/2) - \sin^2(f(n)/2)]^{1/2}h - \sin(f(n))h^2/2 + \mathcal{O}(h^3). \tag{2.22}$$

2.2: A different algorithm, with a smaller truncation error of $\mathcal{O}(h^4)$ and which does not suffer from the $\pm$ complication is

$$f(3) = -f(1) + 2f(2) - \sin(f(2))h^2 + \mathcal{O}(h^4), \tag{2.23}$$

and, in general, for $n \geq 2$

$$f(n+1) = -f(n-1) + 2f(n) - \sin(f(n))h^2 + \mathcal{O}(h^4). \tag{2.24}$$

Show the validity of this algorithm. Hint: compare the Taylor series for $f(n+1)$ and $f(n-1)$.

2.3: Using Eq. (2.20) and implementing a for-loop using Eq. (2.22), numerically find the values of $f(n)$, $n = 1, 2, 3, \ldots, n_{\max}$, and plot $\theta = f(n)$ as a function of time over a time interval of three pendulum periods. Use $\theta_0 = 50°$ (transform into radians). In the same graph, also plot the expression $\theta = \theta_0 \cos(\bar{t})$, which is based on the approximation $\sin(\theta) = \theta$. Try various values of h.

2.4: Repeat Assignment 2.3 by making a for-loop using algorithm (2.24). Compare the difference of that result with the result of part 2.3. A good method is to obtain the absolute value of that difference and plot it on a semilogy plot as a function of θ. The command in MATLAB for absolute value is *abs*.

2.5: (Mathematical) Suppose that you have solved Eq. (2.15) out to an angle $\theta_0 = \theta(\bar{t}_0)$, and now would like to extend the solution by a small angle $\Delta\theta$, i.e., $\theta = \theta_0 + \Delta\theta$. By inserting this expression into Eq. (2.15), and after ignoring terms of order $\mathcal{O}(\Delta\theta^2)$, you find

$$\Delta\theta'' + \cos(\theta_0)\Delta\theta = -\sin(\theta_0). \tag{2.25}$$

Obtain a mathematical solution of Eq. (2.25), by determining the time dependence of $\Delta\theta$ that is compatible with the boundary conditions $\Delta\theta(\bar{t}_0) = 0$, and $\Delta\theta'(\bar{t}_0) = \theta'(\bar{t}_0)$.
Hint: define the new variable $y(\bar{t}) = \Delta\theta(\bar{t}) + \tan(\theta_0)$, and solve the differential equation for y.

2.6: For the oscillations described in Project 1, and based on force diagrams, calculate the tension in the pendulum string of length $\ell = 0.5\,\mathrm{m}$ for a mass $m = 0.25\,\mathrm{kg}$. Make a plot of the tension (units $N = $ Newton) as a function of the angle θ.

2.7: Assume that the tensile strength of the string is $T = 50\,\mathrm{MPa} = 50 \times 10^6\,\mathrm{N/m^2}$ (value which corresponds to nylon string). Calculate the minimum diameter that the string has to have so as not to rupture.

2.8: (Difficult) Assume that the material of the string has some elasticity, and it experiences a small increase in length proportionally to the tension (like Hooke's law for a spring). Derive the correction to the equation of motion (2.15) due to this effect.

General Note: These solutions become really interesting if the initial velocity is not zero. If the initial velocity is directed upward, the pendulum will be thrusted upwards to a certain point where it will stop, and it starts to turn around downward again. Please also note that since the pendulum equation for θ is non-linear, the methods based on a spectral expansion described in Chap. 3 and onward can not be applied directly, other than by an iterative approach.

2.4 The Physics of the Descending Parachute

In this physical problem we follow the descent of a parachute, starting from rest, and keeping in mind that air friction plays a large role. In this initial project we obtain and plot the analytic solution. In a subsequent project we solve the same problem numerically using Euler-Cromer's method.

We take a coordinate system with the y-axis pointing vertically up, the force of gravity $F_g = mg$ pointing straight down, and the force of friction $F_f = k\,v^2$ pointing straight up. Here v is the velocity of the parachute. It is a negative quantity. The parachute is released from the point $y = 0$ with zero velocity. We assume the force of friction to be proportional to the square of the velocity, with the constant

$c = km$ containing the drag coefficient of friction, and the area of the parachute. Using $\mathbf{F}_{total} = m\mathbf{a}$ one obtains the differential equation for v

$$m\frac{dv}{dt} = c\,v^2 - mg. \tag{2.26}$$

After some rearranging one obtains the integral equation

$$\int \frac{dv}{1 - K^2\,v^2} = -\int g\,dt, \tag{2.27}$$

where

$$K = \sqrt{\frac{k}{g}}. \tag{2.28}$$

By using the integral

$$\int \frac{dx}{a + b\,x^2} = \frac{1}{2\sqrt{-a\,b}}\log\left[\frac{a + x\,\sqrt{-a\,b}}{a - x\,\sqrt{-a\,b}}\right], \tag{2.29}$$

with $a = +1$, and $b = -K^2$, and remembering that the initial value of $x = v$ is 0, one obtains

$$\frac{1}{2\,K}\ln\left[\frac{1 + K\,v}{1 - K\,v}\right] = -gt. \tag{2.30}$$

Here ln represents the natural logarithm. Upon introducing the velocity $v_T = 1/K$, one finally obtains from Eq. (2.30)

$$\frac{v}{v_T} = \frac{-1 + e^{-2Kgt}}{1 + e^{-2Kgt}}. \tag{2.31}$$

2.4.1 Assignments

Using the following numbers for the values of

$$c = 12\,\text{kg/m}; \quad m = 100\,\text{kg}; \quad g = 9.8\,\text{m/s}^2,$$

do the assignments proposed as follows.

2.9: Show that v_T is the terminal velocity and show also that $v_T = 1/K$. Show this both by starting from Eq. (2.31) as well as by equating the weight with the air friction.

2.10: Plot v as a function of t (you could use MATLAB).

2.11: Calculate from Eq. (2.31) the time required for the velocity to reach 90% of the terminal velocity. Use pencil and paper.

2.12: Check whether your answer in part 2.11 is compatible with your graph in Problem 2.10.

2.13: (Difficult) If a person strapped to the already open parachute (total mass $M = 100$ kg) jumps from the top of a building that is 60 m high, how long will it take to reach ground, and what will the speed of the parachute be at that time?
  Hint: by doing an integral, $\int_0^t v(t')dt'$ one obtains the distance travelled during the time interval $[0, t]$. Use $\int \tanh(x)dx = \ln(\cosh(x))$.
  Answer: It will take 7.3 s, and the person will hit the ground with a speed of 9.04 m/s $= 20.2$ mi/h.

References

1. W.J. Olver, D.W. Lozier, R.F. Boisvert, C.W. Clark, *NIST Handbook of Mathematical Functions*, National Institute of Standards and Technology, U.S. Department of Commerce (Cambridge University Press, Cambridge, 2010)
2. L.N. Trefethen, *Finite Difference and Spectral Methods for Ordinary and Partial Differential Equations*, Cornell University, Department of Computer Science (Center for Applied Mathematics, Ithaca, 1996)
3. R.J. LeVeque, *Finite Difference Methods for Ordinary and Partial Differential Equations, Steady State and Time Dependent Problems* (SIAM, Philadelphia, 2007)
4. R. Feynman, R. Leighton, M. Sands, *The Feynman Lectures on Physics*, vol. III (Addison Wesley Publishing Co, Reading, 1965)
5. E. Schrödinger, Ann. Phys. **384**, 489–527 (1926)
6. S. Koonin, D. Meredith, *Computational Physics (Fortran version)* (Westview Press, Boulder, 1990)
7. A. Garcia, *Numerical Methods for Physics* (Prentice Hall Inc, Upper Saddle River, 2000)
8. A. Gilat, V. Subramaniam, *Numerical Methods for Engineers and Scientists*, 2nd edn. (Wiley, New York, 2011)
9. B.D. Shizgal, *Spectral Methods in Chemistry and Physics. Applications to Kinetic Theory and Quantum Mechanics* (Springer, Dordrecht, 2015)
10. K.R. Symon, *Mechanics* (Addison Wesley Publishing Company, Reading, 1960)
11. G.B. Arfken, H.J. Weber, *Mathematical Methods for Physicists*, 4th edn. (Academic, San Diego, 1966)
12. M. Abramowitz, I. Stegun (eds.), *Handbook of Mathematical Functions* (Dover, New York, 1972)
13. D.A. McQuarrie, *Mathematical Methods for Scientists and Engineers* (University Science Books, Sausalito, 2003), p. 628
14. L. Verlet, Phys. Rev. **159**, 98 (1967)

Chapter 3
Galerkin and Collocation Methods

Abstract One method to numerically solve an equation consists in expanding the solution in terms of a set of known basis functions. These are the so-called "spectral" methods, where the main emphasis is placed on establishing procedures to obtain the expansion coefficients. In this chapter, we present and compare two such methods: the Galerkin and the Collocation methods, with considerations about the nature of the support points employed by each.

3.1 The Objective and Motivation

The objective of this chapter is to describe the use of the Galerkin and Collocation methods to solve a differential or integral equation. The difference between those methods and the finite difference methods described in Chap. 2 is based on the nature of the errors that occur in each method. As the distance h between successive points in the finite difference method is made smaller the truncation error decreases, but the accumulation of round-off errors increases. If h is made too small, the accumulation of round-off errors overwhelms the truncation error and the result becomes meaningless. At this juncture, the method that expands the solution in a set of basis functions becomes preferable, as is demonstrated by means of numerical examples in Chaps. 6 and 7. This method is the subject of the present and several subsequent chapters, even though it is not directly applicable to nonlinear equations. The reason it is not directly applicable to non-linear equations is because the expansion coefficients themselves also obey non-linear equations. When the number of expansion coefficients is large, those equations become unwieldy to solve. This difficulty does not occur if the non-linear terms are treated iteratively.

3.2 Introduction to Galerkin and Collocation Methods

Assume that the equation to be solved for the function $u(x)$ is

$$\hat{L}u = f, \tag{3.1}$$

© Springer Nature Switzerland AG 2018
G. Rawitscher et al., *An Introductory Guide to Computational Methods for the Solution of Physics Problems*,
https://doi.org/10.1007/978-3-319-42703-4_3

where $\hat{L}$ is a linear operator either in differential or integral form, the function $f(x)$ is given, and the independent variable x is contained in some interval $[a, b]$, that can be either open or closed, or of the form $[0, \infty)$.

A common method to solve for u is to expand it in terms of a complete (but not necessarily orthogonal) set of basis functions $\phi_i(x), i = 1, 2, \ldots, N, \ N + 1, \ldots, \infty$, and solve for the expansion coefficients a_i. However, the expansion has to be truncated at some upper limit $N + 1$, thus introducing an algorithm error. Hence the result, $u^{(N)}$,

$$u^{(N)}(x) = \sum_{i=1}^{N+1} a_i \, \phi_i(x), \quad a \le x \le b \tag{3.2}$$

is only an approximation to the exact solution u. The aim is to minimize the error, called remainder $\mathcal{R}$

$$\hat{L} \, u^{(N)}(x) - f(x) = \mathcal{R}^{(N)}(x). \tag{3.3}$$

The upper limit $N + 1$ of the sum in Eq. (3.2) is chosen in anticipating that the basis functions $\phi_i(x)$ could be polynomials of order $n = i - 1$, in which case the approximant $u^{(N)}(x)$ is a polynomial of order N. In the limit $N \to \infty$, one has $u^{(N)} \to u$ and $\mathcal{R}^{(N)} \to 0$. Please note that at this point x is a continuous variable, and the size of $\mathcal{R}$ may not be uniform (i.e., limited by an upper limit for all x within the interval $[a, b]$), and hence integrals over this interval are employed in order to smooth-out the non-uniformity. However, this is not the case for the collocation methods, where integrals are not performed explicitly, as described below. In the discussions below all the functions involved are assumed to be real.

3.3　The Galerkin Method

In one of the simplest forms of the Galerkin method, the overlap integral $(\chi_i|\mathcal{R})$ over the remainder $\mathcal{R}$

$$(\chi_i|\mathcal{R}) = \int_a^b \chi_i(x)\rho(x)\mathcal{R}(x) \, dx \tag{3.4}$$

over any of the set of auxiliary basis functions $\chi_i(x)$ is considered, and is set to zero. In the integral above ρ is a positive weight function that depends on the type of integration being performed. By multiplying both sides of Eq. (3.3) with χ_j, making use of the expansion (3.2), remembering the linearity of the operator $\hat{L}$, and after integrating the result over the interval $[a, b]$ one finds

$$\sum_{i=1}^{N+1} L_{ji} \, a_i - F_j = \left(\chi_j|\mathcal{R}\right) = 0, \quad j = 1, 2, \ldots, N + 1, \tag{3.5}$$

where

$$L_{ji} = \left(\chi_j | \hat{L} \phi_i \right), \quad \text{and} \quad F_j = \left(\chi_j | f \right). \tag{3.6}$$

Here Eq. (3.5) is a matrix equation, and the whole expansion procedure (3.2) is a discretization in the space of the χ_j functions of the operator $\hat{L}$ acting on ϕ_i. Usually the functions χ_j are replaced by the ϕ_j, and L_{ji} becomes a square matrix, and hopefully it admits an inverse, with not too large a numerical error. That error is described by a condition number C. The Galerkin method is also extensively discussed in sections 1.3 and 1.4 of Shizgal's book [1].

Exercise 1. Take for the operator $\hat{L} = d^2/dx^2$, and consider the function $f = e^x$ for $-1 \leq x \leq 1$. Take for the basis set the functions $\phi_i(x) = \cos(k_i x)$, with $k_i = i\pi$, $i = 0, 1, 2, \ldots$, take for the weight function the value $\rho(x) = 1$, and assume for the upper limit N of the expansion (3.2) some convenient value, like 10 or 20. Also assume that the functions χ_j are identical to the functions ϕ_j.

Assignment:
(a) Calculate the expansion coefficients a_i in Eq. (3.2), by solving Eq. (3.5).
(b) Numerically investigate the rate of convergence of the expansion (3.2) as a function of N. Note that all the integrals can be carried out analytically.
(c) Obtain the analytic solution of Eq. (3.1), and investigate the error of the expansions (3.2) obtained in part (b) as a function of N.

3.3.1 Some Useful Comments

1. Suppose that the ϕ_i are solutions of a part $\bar{L}$ of the operator $\hat{L}$, i.e., $\bar{L} \phi_i = \lambda_i \phi_i$, where λ_i are the discrete bound-state eigenvalues that depend on the appropriate boundary conditions of the ϕ_i. If one replaces the χ_i by the ϕ_i, and if one keeps only one expansion function ϕ_0 then one obtains the perturbation theory formulation that is very common in physics applications, and is described in textbooks on Quantum Mechanics. In this case one finds an improved eigenvalue λ, close to λ_0 by successive iterations, and also finds an improved function ψ that is close to ϕ_0. But we can do much better, as shown below.

2. Consider the case that the ϕ_i, $i = 1, 2, \ldots, N + 1$, are sturmian functions. These functions are eigenfunctions of some operator $\bar{L}$, chosen according to the form of the operator $\hat{L}$. They obey the same boundary conditions as the function u in Eq. (3.1), and are defined for the same fixed energy as the energy of u, as is described in Chapter 11 of Ref. [2]. The eigenvalues multiply an auxiliary potential $\bar{V}$ contained in $\bar{L}$, rather than being energy eigenvalues. If both the set of functions ϕ_i and χ_j are the same appropriately chosen sturmian functions, then the sturmian expansion of $u^{(N)}$ (3.2) may converge very rapidly, i.e., N may be small, and the asymptotic behavior of the expansion will be the same as that of u. This approach is made use of in many applications to physics, and is described in Chap. 11. The main challenge in this case is to obtain a practical method to calculate the sturmian functions and the eigenvalues λ_i [3].

3. A good choice of the set $\{\phi_i\}$ is crucial for the rapid convergence of the expansion (3.2), and theorems relating to the size of N required for a desired accuracy (or smallness of $\mathcal{R}$) will be presented further below for the case of spectral expansions. For the finite element procedure the whole large domain of the independent variable is split into macroscopic smaller segments (called elements or partitions), and for each partition an expansion of the type (3.2) is performed. The requirement that the solution u at the end of one partition must match smoothy onto the solution at the beginning of the next partition is also included in the finite element formalism, and places restrictions on the expansion coefficients in each partition [4].

4. An important feature for the numerical implementation of the Galerkin method is that the choice of the discrete support points in the interval $[a, b]$ is not crucial, other than for the requirement that the integrals be as accurate as possible, if done numerically. For example, equidistant mesh points are needed if Simpson's integration rule is used as described in Section 3.5 (ii) of Ref. [2], and taught in elementary computational methods courses. If the integrals can be done analytically, then of course no choice at all of mesh points is required. This is in contrast to the Collocation method described in the next section, where the choice of mesh points becomes critical.

5. A very useful set of basis functions $\phi_i, i = 1, 2, \ldots, N$ are Lagrange polynomials, each of the same order N. There are various types of Lagrange functions [5]. For each type of such functions a set of N support points ξ_j is defined in the interval $[a, b]$, and each Lagrange function ϕ_i goes through zero at all support points with the exception of ξ_i, where its value is unity. The advantage of these functions is that the integrals defined in Eq. (3.6) or (3.4) can be performed very accurately [5] using Gauss integration methods, requiring only the knowledge of f at ξ_j. These functions are also now used in finite element calculations [6], and an accuracy study is contained in Ref. [4].The Lagrange mesh method is also discussed throughout Shizgal [1], especially in section 3.9, where interpolation and differentiation are described.

3.4 Collocation Method

In this case a choice of support points $\xi_i, i = 1, 2, \ldots, N + 1$ in $[a, b]$ is required. A direct connection with the Galerkin method can be established by choosing the set of functions χ_i that are used in Eqs. (3.5) and (3.6) as Dirac delta functions

$$\chi_i(x) = \delta(x - \xi_i), \tag{3.7}$$

in which case Eq. (3.5) becomes

$$\sum_{i=1}^{N+1} L_{ji}^{(C)} \, a_{i_i}^{(C)} - f(\xi_j) = 0, \quad j = 1, 2, \ldots, N + 1, \tag{3.8}$$

with

$$L_{ji}^{(C)} = [\hat{L}\phi_i]_{\xi_j}. \tag{3.9}$$

The symbol $[\hat{L}\phi_i]_{\xi_j}$ in the equation above means that the function $\hat{L}\phi_i$ is to be evaluated at the point ξ_j. An advantage is that no integrals have to be carried out and once the coefficients a_i are obtained from the solution of the matrix equation (3.8), the value of $u^{(N)}$ can be calculated for any continuous value of x from Eq. (3.2). However, one difficulty is in finding a good method to establish the location and number of support points that are suitable for a given problem. One way to remedy this difficulty is to use special functions that vanish at a given set of mesh points. Here are two examples.

Example 1: The Equidistant Fourier Mesh

For a given value of N, the interval is given by $-\frac{1}{2}N \le x \le \frac{1}{2}N$, the N mesh points are equidistant and are given by

$$\xi_i = i - (1 + N)/2, \quad i = 1, \ldots, N. \tag{3.10}$$

For example, for $N = 4$, x ranges from -2 to 2, the mesh points are

$$\xi_i = -\frac{3}{2}, -\frac{1}{2}, \frac{1}{2}, \frac{3}{2}$$

and the corresponding Lagrange–Fourier functions, according to Ref. [5], section 3, are

$$\phi_i(x) = \frac{\sin(\pi(x - \xi_i))}{N \sin(\frac{\pi}{N}(x - \xi_i))}. \tag{3.11}$$

Please note that these functions are not polynomials in the variable x.

Example 2(a): Lagrange Interpolation Functions

These are polynomials all of the same order $N - 1$

$$\mathscr{L}_i(x) = \prod_{k=1}^{N} \frac{x - \xi_k}{\xi_i - \xi_k}, \quad k \ne i; \quad i = 1, 2, \ldots, N. \tag{3.12}$$

If the mesh points ξ_k are Lobatto points, then in addition to the points ± 1, they are located at the positions of the zeros of the x derivative of the Legendre polynomial P_{N-1} in the domain $[-1, 1]$. Some of the values are given in table 25.6 in Ref. [7].

Example 2(b)

An alternative choice [8] is to use a sequence of different orders of a particular orthogonal polynomial $p(x)$, for example Legendre or Chebyshev. In contrast to Example 2(a) the order of each polynomial is not the same, but increases from 0 to $N - 1$, and $\phi_i(x) = p_{i-1}(x)$, $i = 1, 2, \ldots, N$. The support points ξ_k,

$k = 1, 2, \ldots, N$ are the zeros of the polynomial $p_N(x)$. For the Chebyshev case, they can be obtained by means of simple trigonometric expressions, as described below.

For Chebyshev polynomials $\phi_i(x) = T_{i-1}(x)$, $i = 1, 2, \ldots, N$, the discrete orthogonality is of the form

$$\frac{\pi}{N} \sum_{k=1}^{N} T_n(\xi_k) \, T_m(\xi_k) = \frac{\pi}{2} \delta_{n\,m}(1 + \delta_{0\,n}), \quad n < N, \; m < N, \tag{3.13}$$

where the ξ_k are the zero's of T_N, given in the interval $[-1, 1]$ by

$$\xi_k = \cos\left[\frac{\pi}{N}(k - 1/2)\right], \; k = 1, 2, \ldots, N. \tag{3.14}$$

For Examples 2(a) and 2(b) the support points are not equally spaced, which as we will see during the discussion of spectral methods, give rise to a higher accuracy in the expansion (3.2) than if the points were equally spaced. An important additional feature is the use of the Gauss integration expression

$$\int_a^b \psi(x)\rho(x)dx = \sum_{k=1}^{N} w_k \psi(\xi_k) + E_N(\psi), \tag{3.15}$$

where $\rho(x)$ is the weight function, introduced in Eq. (3.15), and the error is $E_N(\psi)$. If the function ψ is a orthogonal polynomial p_n, then the error $E_N(\psi)$, given in Eq. (3.5.19) of Ref. [2], contains a factor $\int_a^b p_N^2(x)w(x)dx$. As a consequence the relation (3.15) is exact if $\psi(x)$ is a polynomial whose order is not greater than $2N + 1$. The weight functions $\rho(x)$ and the weights w_k depend on each type of orthogonal polynomial $p_n(x)$, and they are listed in table 18.3.1 of Ref. [2] for various orthogonal polynomials. The weights w_k and mesh-points ξ_k are listed on the tables in section 3.5 of Ref. [2] for various types of orthogonal polynomials, and the support points ξ_k are the zeros of the polynomial p_N for the case that they do not include the end points a and b. Detailed values of the weight functions $\rho(x)$ and weight factors w_k for various orthogonal polynomials are given in sections 3.5.15 to 3.5.28 in Ref. [2], and in section 25.4 of Ref. [7]. See also Eqs. (2.58) and (2.58) in Ref. [1], and also in Listing 2.2 how the Gauss–Laguerre quadrature points can be calculated in MATLAB. An example for Legendre polynomials is given in table 3.1.

In order to obtain the coefficients a_i in Eq. (3.5) with the collocation method, one can proceed in two ways: for the first method one uses both for the functions χ_j and the functions ϕ_j and the Lagrange functions $\mathscr{L}_j$ defined in Eq. (3.12). Further, using the vanishing of the $\mathscr{L}_i$ at all mesh points other than ξ_i, together with the Gauss integration expression (3.15) one obtains again Eq. (3.8), with the difference that the basis functions and support points are now well defined. In the second method one can use for both the χ_j and the ϕ_j one of the set of orthogonal polynomials p_{j-1} described in Example 2(b). After using Gauss's quadrature, one obtains

Table 3.1 Mesh points and weights for the 5-point Gauss–Legendre integration formula

$\pm \xi_k$	w_k
0.14887 43389 81631 211	0.29552 42247 14752 870
0.43339 53941 29247 191	0.26926 67193 09996 355
0.67940 95682 99024 406	0.21908 63625 15982 044
0.86506 33666 88984 511	0.14945 13491 50580 593
0.97390 65285 17171 720	0.06667 13443 08688 138

$$\sum_{i=1}^{N+1} M_{ji} a_i = F_j, \tag{3.16}$$

where

$$M_{ji} = \sum_{k=1}^{N+1} w_k \phi_j(\xi_k)[L\phi_i]_{\xi_k}, \tag{3.17}$$

and

$$F_j = \sum_{k=1}^{N+1} w_k \phi_j(\xi_k) f(\xi_k). \tag{3.18}$$

The difference between Eqs. (3.8) and (3.16)–(3.18) is that the former requires the values of the functions at only one support point, line by line, while the latter contain sums over all support points in each line. From the computational point of view, this is an advantage.

3.4.1 Details

As shown before, the Galerkin and Collocation methods are given by $\left(\phi_j | \hat{L}u\right) = \left(\phi_j | f\right)$, and $\left(\delta(x - \xi_j) | \hat{L}u\right) = f(\xi_j)$, respectively, with $j = 1, 2, \ldots, N + 1$. Here $(|)$ is the integral $(\phi_j | \psi) = \int_a^b \phi_j(x)\psi(x)\rho(x)dx$ defined in Eq. (3.4). Due to the linearity of the operator $\hat{L}$, and since $u(x) = \sum_{i=1}^{N+1} a_i^{(G)} \phi_i(x)$, the Galerkin equations lead to matrix equations (3.5) and (3.6) for the $a_i^{(G)}$'s. For the Collocation method $u(x) = \sum_{i=1}^{N+1} a_i^{(C)} \phi_i(x)$ and the equations for the expansion coefficients are given by Eqs. (3.8) and (3.9). The Gauss integration approximation, Eq. (3.15), can also be used extensively, where the weights w_k are given [2, 7] for each choice of basis functions $\{\phi\}$.

The Lagrange functions given by Eq. (3.12) make an excellent basis $\phi_i(x) = \mathscr{L}_i(x)$ for the Galerkin method, since they have the property that they vanish at all mesh points other than ξ_i, and all are polynomials of order $N - 1$. In this case $(\mathscr{L}_j | \psi) = \psi(\xi_j) w_j$ (no summation involved), hence $L_{j\,i}^{(G)} = w_j [\hat{L}\mathscr{L}_i]_{\xi_j}$, and $F_j^{(G)} = w_j f(\xi_j)$. In order to solve Eq. (3.16) for the expansion coefficients a_i the inverse of the matrix M_{ji} is required, but its existence is not guaranteed. An extensive discussion of numerous basis functions is given in Ref. [1], section 3.9.2.

For the implementation of the Collocation method, the following succinct notation is convenient:

$$(\psi) \equiv \begin{pmatrix} \psi(\xi_1) \\ \psi(\xi_2) \\ \vdots \\ \psi(\xi_{N+1}) \end{pmatrix} ; \quad (a) \equiv \begin{pmatrix} a_1 \\ a_2 \\ \vdots \\ a_{N+1} \end{pmatrix}. \tag{3.19}$$

Here a parenthesis around a function ψ means a column of the values of ψ evaluated at the support points $\psi(\xi_i)$, and a column of expansion coefficients a_i is denoted as (a). This notation is convenient for subsequent matrix manipulations, because (a) represents a matrix $(N + 1, 1)$. A superscript ψ in (a^ψ) denotes that these $a's$ correspond to the expansion of the function ψ. The superscripts (N) as well as (C) for the approximate function $u^{(N)}(x)$ will be dropped in what follows since the difference between $u^{(N)}(x)$ and $u(x)$ will be considered in a later chapter.

With this notation, the description of the Collocation method is as follows. Given a function $u(x)$, the coefficients $a_i^{(u)}$ of the expansion

$$\sum_{i=1}^{N+1} a_i^{(u)} \phi_i(x) = u^{(N)}(x) \tag{3.20}$$

can be expressed through the matrix relationship

$$(u) = C (a^u), \tag{3.21}$$

where the elements C_{ki} of the $(N + 1) \times (N + 1)$ matrix C are given by $\phi_i(\xi_k)$. This result is obtained by rewriting the expansion $N + 1$ times, once for each value of $x = \xi_k$ with $k = 1, 2, \ldots, N + 1$, thus obtaining a set of linear equations for the coefficients $a_i^{(u)}$. The coefficients then can be obtained in terms of the $u(\xi_k)$ either by solving directly Eq. (3.20), or by obtaining the inverse of C

$$(a^u) = C^{-1}(u). \tag{3.22}$$

Equations (3.21) and (3.22) appear in the work of Curtis and Clenshaw [9] and take on a specially important role for the collocation method. Explicit expressions of the matrices for both C and C^{-1} by considering a Chebyshev polynomial expansion with $N + 1$ terms is given in Appendix A by means of the function "C_CM1.m". For expansions in other types of polynomials, the matrices for C have to be set-up as described above, and to find C^{-1} the inverse has to be computed explicitly, as described above. The discrete orthogonality property given by Eq. (3.24) of these polynomials can be very helpful in order to find C^{-1} without solving Eq. (3.20).

If the functions $\phi_i(x)$ are orthogonal polynomials $p_n(x)$, with $n = i - 1$, for $n = 0, 1, 2, \ldots, N$ (the lower case p stands for any of the orthogonal polynomials of degree n and is not to be confused with Legendre polynomials P_n) then they obey the orthogonality condition

$$(p_n | p_m) = h_n \, \delta_{n \, m}, \tag{3.23}$$

where the constants h_n depend on the normalization of the functions p_n, and are also listed in Refs. [2, 7]. Based on Eq. (3.15), the discrete form of Eq. (3.23) is

$$\sum_{k=1}^{N+1} p_n(\xi_k) \, p_m(\xi_k) w_k = h_n \, \delta_{n \, m} \, , \, n + m \le 2N + 1, \tag{3.24}$$

where the ξ_k are $N + 1$ mesh points in the interval $[a, b]$ [2, 7].

If one still assumes that the functions ϕ_i are orthogonal polynomials, the inverse of the matrix C can also be obtained by writing Eq. (3.20) repeatedly for x replaced by $\xi_k, k = 1, 2, \ldots, N + 1$, multiplying each of these equations by $w_k \phi_i(\xi_k)$, summing over k and making use of Eq. (3.24). One finds

$$a_j = \sum_{k=1}^{N+1} \frac{w_k}{h_k} \phi_j(\xi_k) u^{(N)}(\xi_k), \tag{3.25}$$

which shows that the elements $(C^{-1})_{j\,k}$ of the inverse of the matrix C are given by $(w_k/h_k)\phi_j(\xi_k)$.

With this notation the collocation method $\hat{L}u|_{\xi_k} = \sum_{i=1}^{N+1} \hat{L}\phi_i(\xi_k)a_i^{(u)} = f(\xi_k)$, $k = 1, 2, \ldots, N + 1$, can be developed further by expanding each function $\hat{L}\phi_i(x)$ in terms of the basis functions $\phi_t(x)$, $t = 1, 2, \ldots, T$, so that

$$\hat{L}\phi_i(x) = \sum_{t=1}^{T} b_{t\,i} \, \phi_t(x). \tag{3.26}$$

With that notation $\hat{L}u(x) = \sum_t^T \phi_t(x) \sum_i^{N+1} b_{t\,i} \, a_i^{(u)}$. The second sum can be expressed in matrix form

$$\sum_i^{N+1} b_{t\,i} \, a_i^{(u)} = B_t(a^{(u)}) \tag{3.27}$$

where B_t is a $(1, N + 1)$ matrix whose row contains $b_{t\,i}, i = 1, 2, \ldots, N + 1$. Equating $\sum_t^T \phi_t(x) \sum_i^{N+1} b_{t\,i} \, a_i^{(u)}$ to $f(x) = \sum_{i=1}^{N+1} a_t^{(f)} \phi_t(x)$, assuming that the upper limit T in Eq. (3.26) is the same as $N + 1$, and setting equal to each other the coefficients of the same functions $\phi_t(x)$ on each side of this equation, one obtains

$$B(a^{(u)}) = (a^{(f)}), \tag{3.28}$$

where the elements of the matrix B are given by

$$(B)_{t\,i} = b_{t\,i}. \tag{3.29}$$

If $T \neq N + 1$, then the matrix B will not be square, but will be of dimension $(T, N + 1)$, where T denotes the number of lines and $N + 1$ denotes the number of columns. If $T = N + 1$, and if the inverse of the matrix B exists, then the solution for the expansion coefficients $a_i^{(u)}$ is given by

$$(a^{(u)}) = B^{-1}(a^{(f)}). \tag{3.30}$$

Taking for the values of x in the equations above the mesh point values ξ_k, $k = 1, 2, \ldots, N + 1$, and remembering that $(u) = C(a^{(u)})$ and $(f) = C(a^{(f)})$, Eq. (3.30) can be written in terms of the column vectors (u) and (f) as

$$(u) = C^{-1}B^{-1}C\ (f). \tag{3.31}$$

Given a general function $v(x)$, $x \in [a, b]$, it can be seen from the left hand side of Eq. (3.28) that the expansion coefficients of $(\hat{L}v)$ are $B(a^{(v)})$, and hence

$$(\hat{L}v) = BC^{-1}(v). \tag{3.32}$$

Please note that the matrices B and C do not depend on the function v, but only on the choice of the type of polynomials p_i. For example, if $\hat{L} = d/dx$ is the differentiation operator, then BC^{-1} is the differentiation matrix in the Collocation representation. An explicit example for this differentiation operator is given below. If $\hat{L} = \int_a^x dx$ is the integral operator, then the corresponding matrix will be given in Chap. 5.

3.4.1.1 Example for the Differentiation Operator

In this first example the function to be differentiated with respect to r is

$$u(r) = \frac{1}{a}e^{(r-R)/a}/[1 + e^{(r-R)/a}]^2 \tag{3.33}$$

and the derivative with respect to r is given by

$$f(r) = (1/a^2)(1 + y)^{-2}[1 - 2y/(1 + y)]; \quad y = e^{(r-R)/a}. \tag{3.34}$$

Fig. 3.1 The test functions u
and $f = du/dr$, as
calculated from their
analytical expressions,
Eqs. (3.33) and (3.34). The
symbols are located at the
Chebyshev support points
for the number of Chebyshev
expansion functions
$N + 1 = 51$

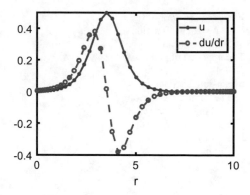

A plot of these functions is given in Fig. 3.1.

$$u(r) = \sum_{n=1}^{N+1} a_n \, T_{n-1}(x), \quad -1 \le x \le 1,$$

where u is known, and f (now denoted as the column vector (f)) is calculated at the Chebyshev support points by means of the matrix representation of the differentiation operator $D = B \, C^{-1}$

$$(f) = D(u). \tag{3.35}$$

The $(N + 1) \times (N + 1)$ matrices C^{-1} and B are composed of values of the Chebyshev polynomials and their derivatives at the Chebyshev support points $\xi_1, \xi_2, \ldots, \xi_{N+1}$. In particular, B and C are composed of $N + 1$ columns $k = 1, 2, \ldots, N + 1$, each containing the derivative of T_k (for B) or the values of T_k (for C). The inverse can be obtained, as is explained in detail in Chap. 5. An equation similar to Eq. (3.35), but more general, is also given by Eqs. (1.67), (3.131) and (3.138) in Ref. [1], together with numerical examples of the accuracy obtained.

In the numerical calculation the matrix C^{-1} is given by the MATLAB function $[C, CM1, xz] = C\_CM1(N)$, reproduced in Appendix B, where it is denoted by $CM1$, the support points in the interval $[-1, 1]$ are given by the column vector (xz), and the matrix B is given by $CHder1$ which can be called from the function $[CH, CHder1, CHder2] = der2CHEB(xz)$. This function is also listed in Appendix B, and a listing of the program that calculates the function f is listed as "Program Deriv_u.m" in Appendix B.

The numerical results for f in the interval $0 \le r \le 10$, and for values of $R = 3.5$ and $a = 0.5$, were calculated for two cases with $N = 25$ and 50. The achieved accuracies are illustrated in Fig. 3.2. It is worth noting that doubling the number of expansion polynomials increases the accuracy by three orders of magnitude.

A comparison with the accuracy obtained in Output 11, on p. 56 of Ref. [10] for the derivative of the function

Fig. 3.2 The error of the derivative of the function u, as calculated from Eq. (3.35). The upper set of points uses $N + 1 = 26$ Chebyshev expansion functions and let to an error of the order of 10^{-2}. For the lower set of points $N + 1 = 51$, the error is of order 10^{-5}, as shown in the figure

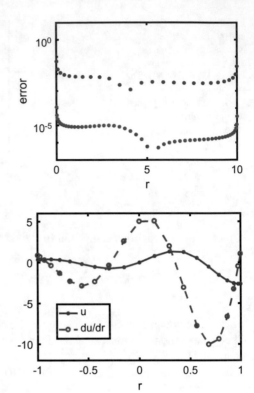

Fig. 3.3 The test functions $u = \sin(5r)\exp(r)$ and $f = du/dr$, as calculated from their analytical expressions. The symbols are located at the Chebyshev support points for the number of Chebyshev expansion functions $N + 1 = 21$

$$u = e^r \sin(5r), \; -1 \le r \le 1, \tag{3.36}$$

using 20 Chebyshev polynomials, but at a different set of support points that include the points ± 1, shows that the method here described gives the same accuracy using 21 Chebyshev polynomials, of the order of 10^{-10}. The reader is forewarned that the support points used in the present formulation do not include the end-points ± 1 as is explained in Chaps. 5 and 6. This result is displayed in Figs. 3.3 and 3.4.

The result of Fig. 3.4, when compared with Fig. 3.2, shows that the accuracy depends both on the nature of the function whose derivative is being calculated, as well as on the number of expansion functions. This point is discussed in detail in Chap. 4.

Using the Lagrange basis set, the Galerkin method can also be developed along these lines, and is well suited for the calculation of bound eigenstates of a Schrödinger equation. In this case $\hat{L} = \hat{T} + V - E$, where $\hat{T} = -d^2/dx^2$, and one can rewrite $(\hat{T} + V - E)u = 0$ as

$$\sum_{i=1}^{N} (T_{ji} + V_{ji})a_i = \sum_{i=1}^{N} E_{ji}a_i = E_j a_j. \tag{3.37}$$

Fig. 3.4 The error of the first and second derivatives of the function $u = \sin(5r)\exp(r)$, as calculated from Eq. (3.35). The number of Chebyshev expansion functions is $N + 1 = 21$

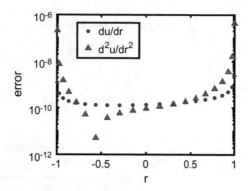

The last step in Eq. (3.37) is due to the fact that the matrix E_{ji} is diagonal, and hence the eigenvalues of the matrix $(T_{ji} + V_{ji})$ are the energy eigenvalues E_j. In order to properly implement this scheme, it is important that the basis functions obey the correct boundary conditions, i.e., vanish as $x \to \pm\infty$. Laguerre polynomials might be a good choice.

In summary, here we have presented the two main forms of the solutions of Eq. (3.1) in terms of expansions into a set of basis functions, namely the Galerkin and the Collocation methods. We still have not mentioned the rate of convergence of such expansions, which will be the subject of Chap. 4. It is to be noted, however, that if the operator $\hat{L}$ is non-linear, as was the case for the pendulum equation presented in Chap. 2, then such expansions cannot be used, and have to be supplemented by iteration procedures, as illustrated by means of an example in Chap. 8. If $\hat{L}$ is linear, however, the accuracy of the solution of Eq. (3.1) can be higher than if the finite difference method is used, as will be demonstrated in Chaps. 6 and 7.

3.4.2 Advantage of a Non-equispaced Mesh

Collocation calculations based on support points that are not equispaced tend to give a higher accuracy than if the support points are equispaced, as observed by Lanczos as early as 1938 [11, 12]. A very enlightening proof is given in Chap. 5, Eq. (2.8), on p. 13 of Ref. [10]. The same is true for interpolation procedures. The proof is based on a comparison with electric potentials due to a charge distribution, but will not be presented here. Nevertheless, because this result is so important, the example given on p. 24 of Ref. [10] will be reproduced here.

In this example the function being interpolated by means of a polynomial is

$$u(x) = 1/(1 + 16\,x^2), \quad x \in [-1, 1].$$

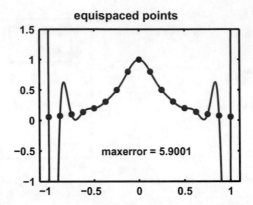

Fig. 3.5 In the plot, $N = 16 + 1$ equispaced points for $x = [-1, 1]$ interpolate the function $u(x) = 1/(1 + 16\,x^2)$. The program is Program9 of Trefethen. It uses $p = polyfit(x, u, N)$ to obtain the coefficients of the polynomial $a_0 + a_1 x + \cdots a_N\, x^N$. The polynomial is then evaluated at all points xx by means of $pp = polyval(p, xx)$. The result $pp(x)$ is plotted by means of the solid line; the max error is given below the curve

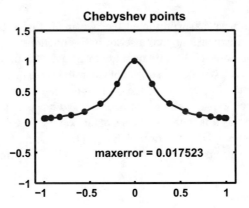

Fig. 3.6 Non-equispaced Chebyshev points in the same interval $[-1, 1]$, interpolating the same function as in Fig. 3.5, but using $N = 16 + 1$. The Chebyshev non-equispaced points in $[-1, 1]$ are given by $x_j = \cos(j\pi/N)$, $j = 0, 1, \ldots, N$. These points include -1 and $+1$. These support points are not the zeros of a Chebyshev function $T_{17}(x)$, which do not include -1 and $+1$, and are spaced slightly, differently from the ones above

The procedure consists in approximating the function $u(x)$ by a polynomial which is fitted to all the discrete values of the function, evaluated at the discrete support points. Hence the method is spectral. However, in one case the points are equidistant, in the other they are clustered according to the Chebyshev method, described in Chap. 5. The results are displayed in Figs. 3.5 and 3.6.

References

1. B.D. Shizgal, *Spectral Methods in Chemistry and Physics. Applications to Kinetic Theory and Quantum Mechanics* (Springer, Dordrecht, 2015)
2. W.J. Olver, D.W. Lozier, R.F. Boisvert, C.W. Clark, *NIST Handbook of Mathematical Functions*, National Institute of Standards and Technology, U.S. Department of Commerce (Cambridge University Press, Cambridge, 2010)
3. G.H. Rawitscher, Positive energy Weinberg states for the solution of scattering problems. Phys. Rev. C **25**, 2196–2213 (1982); G. Rawitscher, Iterative solution of integral equations on a basis of positive energy Sturmian functions. Phys. Rev. E **85**, 026701 (2012); M.J. Ambrosio, F.D. Colavecchia, G. Gasaneo, D.M. Mitnik, L.U. Ancarani, Double ionization of helium by fast electrons with the generalized Sturmian functions method. J. Phys. B: At. Mol. Opt. Phys. **48**(5), 055204 (2015)
4. J. Power, G. Rawitscher, Phys. Rev. E **86**, 066707 (2012)
5. D. Baye, Phys. Stat. Sol. **243**, 1095–1109 (2006); D. Baye, The Lagrange-mesh method. Phys. Rep. **565**, 1–107 (2015)
6. T.N. Rescigno, C.W. McCurdy, Phys. Rev. A **62**, 032706 (2000)
7. M. Abramowitz, I. Stegun (eds.), *Handbook of Mathematical Functions* (Dover, New York, 1972)
8. A. Deloff, Ann. Phys. (NY) **322**, 1373–1419 (2007)
9. C.C. Clenshaw, A.R. Curtis, Numer. Math. **2**, 197 (1960)
10. L.N. Trefethen, *Spectral Methods in MATLAB* (SIAM, Philadelphia, 2000)
11. P.Y.P. Chen, Appl. Math. **7**, 927 (2016). https://doi.org/10.4236/am.2016.79083
12. C. Lanczos, J. Math. Phys. **17**, 123 (1938). https://doi.org/10.1002/sapm1938171123

Chapter 4
Convergence of Spectral Approximations

Abstract In this chapter we discuss by means of general theorems the rate of convergence and accuracy of spectral methods. These are the methods so called "spectral" methods that consist in expanding the solution to a particular problem in terms of a set of basis functions. We initially present theorems about the convergence of Fourier transforms, alongside with the accuracy of a Fourier spectral differentiation. Next, we present theorems concerning the calculation of the interpolation error and the determination of the set of functions that gives rise to the best interpolation in such methods. In last section, we present assignments in order to analyze the rate of convergence and the error of various sets of basis functions in an expansion.

4.1 The Objective and Motivation

In this chapter we present theorems about the rapidity of convergence of the expansion of a given function $u(x)$ in terms of various types of basis sets. We also present the construction of spectral derivative matrices suitable to solve a differential equation, and discuss the expected accuracy. The method consists in approximating the function u by means of another function v which passes through all the mesh points (also called support points) located in a specific radial domain. For a given set of support points, u and v coincide at the support points. But in the region between support points, u and v differ. Estimating the size of this difference (which is an algorithm error) is the object of the present chapter. The function v is obtained by means of the expansion of u in a basis set, and has the property that its derivatives can be obtained analytically. If the expansion basis set is composed of polynomials, then the approximant v is also a polynomial of a given order, which is unique. In Chap. 3 we mentioned that for the collocation method the expansion of functions in terms of basis sets using non-equispaced points, the precision is much higher than that with equispaced points. That conclusion, based on a particular example, is a general property, as is shown by Trefethen in his book [1], and plays a key role in the present chapter. Many basis functions used for such spectral expansions are presented in section 1.2.1 of Ref. [2], with comments on their convergence and references to many of the investigations on this topic.

© Springer Nature Switzerland AG 2018
G. Rawitscher et al., *An Introductory Guide to Computational Methods for the Solution of Physics Problems*,
https://doi.org/10.1007/978-3-319-42703-4_4

4.2   Fourier Expansions

Theorems 1, 2 and 3 are based on Ref. [1], Chapter 3, while theorems 4 and 5 are based on Ref. [3].

Given a continuous function $u(x)$ and the discrete set of mesh points x_m, $m = 1, 2, \ldots, N$, then a function $v(x)$ based on these mesh points is defined such that at the mesh points

$$u(x_m) = v(x_m) \equiv v_m , \quad m = 1, 2, \ldots, N. \tag{4.1}$$

The Fourier transforms of u and v are

$$\hat{u}(k) = \int_{-\infty}^{\infty} e^{-ikx} u(x)dx,$$

$$\hat{v}(k) = h \sum_{j=1}^{N} e^{-ik\,x_j} v_j, \quad k = -\frac{N}{2} + 1, \ldots, \frac{N}{2}. \tag{4.2}$$

The inverse transforms are

$$u(x) = \frac{1}{2\pi} \int_{-\infty}^{\infty} e^{ikx} \hat{u}(k)dk,$$

$$v_j = \frac{1}{2\pi} \sum_{k=-N/2+1}^{N/2} e^{ik\,x_j} \hat{v}_k, \quad j = 1, \ldots, N. \tag{4.3}$$

Theorem 1 (Convergence of a Fourier Transform)

(a) If u has p − 1 continuous derivatives in $L^2(\mathbb{R})$, and a pth derivative of bounded value, then

$$\hat{u}(k) = O(|k|^{-p-1}) \text{ as } k \to \infty. \tag{4.4}$$

(b) If u has ∞ many continuous derivatives in $L^2(\mathbb{R})$, then

$$\hat{u}(k) = O(|k|^{-m}) \text{ as } k \to \infty \tag{4.5}$$

for every m > 0. The latter means that the convergence is super-algebraic.

Example 1:

$$u(x) = \pi e^{-\sigma|x|},$$

$$\hat{u}(k) = \frac{\sigma}{k^2 + \sigma^2}.$$

Here $\hat{u}(k)$ decays algebraically because $p = 1$ (note the absolute value in the exponent).

Example 2:

$$u(x) = e^{-x^2/2\sigma^2},$$
$$\hat{u}(k) = \sigma\sqrt{\pi/2}\, e^{-\sigma^2 k^2/2}.$$

The Fourier transform decays super-algebraically because u has an infinite number of derivatives.

Theorem 2 (Convergence of a Discrete Fourier Transform)

Let $u \in L^2(\mathbb{R})$, and let v be the grid function defined on $h\mathbb{Z}$ by $v_j = u(x_j)$ (Note that $h\mathbb{Z}$ has an infinite number of discrete points separated by the distance h, going from $-\infty$ to $+\infty$.). Then for all $k \in [-\pi/h, \pi/h]$

(a) If u has $p-1$ continuous derivatives in $L^2(\mathbb{R})$, and a pth derivative of bounded value, $(p \geq 1)$ then

$$|\hat{v}(k) - \hat{u}(k)| = O(h^{p+1}) \text{ as } h \to 0. \tag{4.6}$$

(b) If u has ∞ many continuous derivatives in $L^2(\mathbb{R})$, then

$$|\hat{v}(k) - \hat{u}(k)| = O(h^m) \text{ as } h \to 0 \text{ for every } m > 0, \tag{4.7}$$

i.e., the expansion converges super-algebraically.

Theorem 3 (Accuracy of a Fourier Spectral Differentiation)

Let $u \in L^2(\mathbb{R})$ have a vth derivative $(v \geq 1)$ of bounded variation, and let w be the v' the spectral derivative of u on the grid $h\mathbb{Z}$. Hence, for all $x \in h\mathbb{Z}$ the following holds uniformly:

(a) If u has $p-1$ continuous derivatives in $L^2(\mathbb{R})$ for some $p \geq v+1$, and a pth derivative of bounded value, $(p \geq 1)$ then

$$|w_j - u^{(v)}(x_j)| = O(h^{p-v}) \text{ as } h \to 0. \tag{4.8}$$

(b) If u has ∞ many continuous derivatives in $L^2(\mathbb{R})$, then

$$|w_j - u^{(v)}(x_j)| = O(h^m) \text{ as } h \to 0 \text{ for every } m > 0 \tag{4.9}$$

(i.e., the convergence is super-algebraic).

4.3   Fourier Spectral Differentiation on Bounded Periodic Grids

For periodic functions whose discrete support points are defined on a finite, evenly spaced grid $x_1 = h$, $x_2 = 2h$, $x_{N/2} = \pi, \ldots, x_N = 2\pi$, where $h = 2\pi/N$, and N is even, the Fourier transforms are given by

$$\hat{v}_k = h \sum_{j=1}^{N} e^{-ikx_j} v_j, \ k = -\frac{N}{2} + 1, \ldots, \frac{N}{2}. \tag{4.10}$$

The function $v(x)$ based on these mesh points is such that at the mesh points it coincides with $u(x)$, i.e., according to Eq. (4.1) $v_m = u(x_m) = v(x_m)$. This function has an inverse Fourier transform, which at the mesh points is given by

$$v_j = \frac{1}{2\pi} \sum_{k=-N/2}^{N/2} {}'' e^{ikx_j} \hat{v}_k \, , \ j = 1, 2, \ldots, N, \tag{4.11}$$

where the " means that the terms with $k = \pm N/2$ are multiplied by $1/2$, and the periodicity property sets $\hat{v}_{-N/2} = \hat{v}_{N/2}$. The corresponding band-limited interpolant, valid for all values of x, is

$$p(x) = \frac{1}{2\pi} \sum_{k=-N/2}^{N/2} {}'' e^{ikx} \hat{v}_k \, , \ x \in [0, 2\pi]. \tag{4.12}$$

Based on these results, the band limited interpolant to a delta function $\delta(x)$ is the *periodic sinc function* S_N

$$S_N(x) = \frac{\sin(\pi x/h)}{(2\pi/h) \, \tan(x/2)}, \tag{4.13}$$

given by Eq. (3.7) of Ref. [1]. This function is equal to 1 at points $x = 0$ and 2π, and vanishes at all intermediate discrete points $x_n = nh, \ n = 1, 2, \ldots, N - 1$. This function can be used to build up an interpolant to the function v

$$p(x) = \sum_{m=1}^{N} v_m S_N(x - x_m). \tag{4.14}$$

This is the basic result for Fourier-transforms of periodic functions. (Please notice that here $p(x)$ is not a polynomial.) The derivatives dp/dx at the support points x_m can be obtained analytically in terms of the derivatives of the functions $S_N(x - x_m)$. If $y = (x - x_m)$ one finds

$$S_N'(y) = \frac{dS_N(y)}{dy} = \frac{1}{2} \frac{\cos(\pi y/h)}{\tan(y/2)} - \frac{\sin(\pi y/h)}{(4\pi/h) \sin^2(y/2)}. \tag{4.15}$$

From the above result one finds that

$$S_N'(-y) = -S_N'(y), \tag{4.16}$$

and that

$$S'_N(x_n - x_m) = \frac{\cos[\pi(n-m)]}{2\tan[(n-m)\pi/N]}. \tag{4.17}$$

If $n - m$ is denoted as j, the final result in agreement with Eq. (3.9) of Ref. [1] is

$$S'_N = \frac{(-)^j}{2\tan(j\pi/N)}, \quad j = n - m. \tag{4.18}$$

The derivative matrix operator, denoted as D_N, when acting on the column vector of the values at the support points of the function u, obtains the column vector of the values of the derivative of u at the same support points,

$$\begin{pmatrix} u'(\xi_1) \\ u'(\xi_2) \\ \vdots \\ u'(\xi_N) \end{pmatrix} = D_N \begin{pmatrix} u(\xi_1) \\ u(\xi_2) \\ \vdots \\ u(\xi_N) \end{pmatrix}. \tag{4.19}$$

In Chap. 3 such a matrix appeared in Eq. (3.32), where it was expressed in the general form CBC^{-1}, where the matrix B contains the expansion coefficients of the operator $\hat{L}$ acting on any of the expansion functions $\phi_i, i = 1, 2, \ldots, N$. In the present procedure the function u is expanded into the functions ϕ_i, and the derivative of u is obtained by using the same expansion, and replacing the ϕ_i by the analytic expressions of the derivatives of the ϕ_i. In the present case the functions $\phi_i = S_N(x - x_i)$ are not polynomials, and their derivatives given by Eq. (4.17) can be expressed in terms of sines and cosines. An explicit expression for D_N is given by

$$D_N =$$

0	$\frac{1}{2}\cot(\frac{1h}{2})$	$-\frac{1}{2}\cot(\frac{2h}{2})$	$\frac{1}{2}\cot(\frac{3h}{2})$	$\cdots$	
$-\frac{1}{2}\cot(\frac{1h}{2})$	0	$\frac{1}{2}\cot(\frac{1h}{2})$	$-\frac{1}{2}\cot(\frac{2h}{2})$	$\cdots$	
$\frac{1}{2}\cot(\frac{2h}{2})$	$-\frac{1}{2}\cot(\frac{1h}{2})$	0	$\frac{1}{2}\cot(\frac{1h}{2})$	$\cdots$	
$-\frac{1}{2}\cot(\frac{3h}{2})$	$\frac{1}{2}\cot(\frac{2h}{2})$	$-\frac{1}{2}\cot(\frac{1h}{2})$	$\ddots$	$\ddots$	$\ddots$
$\vdots$	$\vdots$	$\vdots$		$\ddots$	
$-\frac{1}{2}\cot(\frac{(N-1)h}{2})$	$\frac{1}{2}\cot(\frac{(N-2)h}{2})$	$-\frac{1}{2}\cot(\frac{(N-3)h}{2})$	$\cdots$		$-\frac{1}{2}\cot(\frac{1h}{2})$

$$\cdots \begin{pmatrix} \frac{1}{2}\cot(\frac{(N-1)h}{2}) \\ -\frac{1}{2}\cot(\frac{(N-2)h}{2}) \\ \frac{1}{2}\cot(\frac{(N-3)h}{2}) \\ \vdots \\ \frac{1}{2}\cot(\frac{1h}{2}) \\ 0 \end{pmatrix} \tag{4.20}$$

For small values of the argument, using $\cot(x) \simeq 1/x$, the entries in the matrix above are of order $(1/ih)$. That value is preferable to the case where the ϕ_i are polynomials of the order N, for which the derivative matrix has larger values for large values of i, as will be discussed in Chap. 8: " The phase -amplitude method".

An example is given in Ref. [1] for the differentiation of the function $\exp[\sin(x)]$ for x in the interval $[0, 2\pi]$. The basis functions are given by the periodic $\sin c$ function S_N, Eq. (4.13), and the derivatives given by Eq. (4.15), where N is the number of equispaced points in the interval $[0, 2\pi]$, and h is the distance between them. The expansion of a periodic function $u(x)$ is given by $u^{(N)}(x) = \sum_{m=1}^{N} v_m S_N(x - x_m)$, Eq. (4.14), as described in Ref. [1], p. 21. For the example of $u(x) = \exp[\sin(x)]$ the accuracy of the derivative of this periodic function is excellent, of the order of 10^{-12} for $N = 24$, while the accuracy of the derivative of a "hat" function is very poor (please see Output 4 in Ref. [1]). By comparison, an expansion of the derivative of this function in terms of Chebyshev polynomials, according to Eq. (3.35) with $N = 25$, using our Chebyshev mesh of support points, Eq. (3.14), resulted in an error at both $x \simeq 0$ and $x = 2\pi$ of order 3×10^{-5}. That is much larger than the error 10^{-12} stated by Trefethen, and shows that the non-polynomial expansion functions are much better than the polynomial-based expansion functions in order to calculate derivatives for periodic functions. Spectral differentiation methods can be found in additional references, such as in Ref. [4]. In Appendix B a MATLAB implementation of our algorithm of the derivative matrix is given.

Trefethen also gives the elements of the second order derivative of a periodic function, with the result $S_N''(x_j) = -\pi^2/(3h^2) - 1/6$ for $j = 0$ (Mod N), and $S_N''(x_j) = -(-1)^j/[2\sin^2(jh/2)]$ for $j \neq 0$ (Mod N). For small values of h, these matrix elements for small values of j are of the order of h^{-2}.

4.4 Convergence of a Polynomial Approximation to a Function

The expansions into a set of polynomials amount to creating a polynomial $p_N(x)$ of order N which, for a given set of mesh points ξ_i, $i = 1, 2, \ldots, N$ has values that at the mesh points are equal to the function being expanded, i.e., $p_N(\xi_i) = f(\xi_i)$, $i = 1, 2, \ldots, N$. For example an expansion of a function $f(x)$ into a truncated set of $N + 1$ Chebyshev polynomials $T_n(x)$, giving rise to a polynomial of order N, is given by

$$p_N(x) = \sum_{j=1}^{N+1} a_j T_{j-1}(x), \quad -1 \leq x \leq 1. \tag{4.21}$$

It is of interest to know how well p_N approaches f at points x other than the mesh points. That knowledge determines the interpolation error. Two theorems are relevant

[3]. The second theorem shows that the Chebyshev mesh points give rise to the best interpolation.

Theorem 4 (Cauchy Interpolation Error Theorem)

Let $f(x)$ be a function sufficiently smooth so that it has at least $N + 1$ continuous derivatives on the interval $[-1, 1]$, and let p_N be its Lagrangian interpolant of degree N for $x \in [-1, 1]$. Then the upper bound of the interpolation error is given by

$$f(x) - P_N(x) \leq \frac{1}{(N+1)!} f^{(N+1)}(\bar{x}) \prod_{i=1}^{N} (x - \xi_i) \tag{4.22}$$

for some $\bar{x} \in [-1, 1]$.

Deloff [3] notices that in order to minimize the interpolation error $f(x) - p_N(x)$ for any function f the product term in Eq. (4.22) should be minimized. This product term has the monic character, i.e., the coefficient of the highest power of x is unity. The theorem below states that the best choice of the ξ_i are the Chebyshev support points.

Theorem 5 (Chebyshev Minimal Amplitude Theorem)

Out of all monic polynomials of degree N, the unique polynomial which has the smallest maximum on $[-1, 1]$ is the monic Chebyshev polynomial $T_N / 2^{N-1}$, i.e., all monic polynomials $Q_N(x)$ satisfy he inequality

$$\max |Q_N(x)| \geq \max |T_N(x)/2^{N-1}| = 1/2^{N-1}, \tag{4.23}$$

for all $x \in [-1, 1]$.

Comments: The same polynomial p_N can be obtained either by a sum over Lagrange functions times coefficients or a sum over Chebyshev (or other orthogonal) polynomials, so that the maximum power of x is N. Hence, according to Theorem 4, for a given N, the error $f(x) - p_N(x)$ is less or equal to a constant factor times the monic polynomial $\prod_{i=1}^{N}(x - \xi_i)$. The coefficient of the highest order term x^N of the Chebyshev polynomial $T_N(x)$ is 2^{N-1}, hence $T_N(x)/2^{N-1}$ is a monic polynomial. According to Theorem 2 the monic polynomial that has the minimum amplitude is $T_N(x)/2^{N-1}$, if the $\{\xi_i\}_N$ are the zeros of T_{N+1}. Further, according to Trefethen, Chap. 5, these amplitudes are all uniform in x, i.e., they are all bounded by the same superating constant. Hence, the expansion of a function in terms of Chebyshev polynomials gives not only the smallest error, but one that is also uniform. This is not the case when equispaced points are used in the construction of the interpolant $p_N(x)$, as is illustrated at the end of Chap. 3.

The rule of thumb for expansions into Chebyshev polynomials is that the truncation error in the expansion after $N + 1$ terms is closely given by the magnitude of the next coefficient of the expansion, a_{N+2}. This is also based on Theorem 2 in Chapter XI, section 11.7 of the book by Luke [5]

$$|p_N(x) - \bar{f}(x)| \simeq a_{N+1} T_{N+1}(x) \left[1 + 2x \, a_{N+1}/a_{N+2} \right]. \tag{4.24}$$

For practical applications it can be assumed that

$$|p_N(x) - \bar{f}(x)| \leq |a_{N+1}|. \tag{4.25}$$

This property enables one to pre-assign an accuracy requirement *tol* for the expansion (4.21). Either, for a given value of N, the size of the partition of r within which the function $f(r)$ is expanded can be determined, or for a given size of the partition, the value of N can be determined such that the sum of the absolute values of the three last expansion coefficients a_{N-2}, a_{N-1} and a_N is less than the value of *tol*. An example is given in Figs. 5.5 and 5.6 in Chap. 5 for the expansion of $\exp(x)$ in the interval $[-1, 1]$. The figures show that the error ε_N divided by a_{N+2} of the expansion $\exp(x) = \sum_{k=1}^{N+1} a_k T_{k-1}(x) + \varepsilon_N$ (truncated at $N + 1$) is in magnitude less than $\simeq 1$, and is uniform in x.

For functions that are periodic, a Fourier expansion may converge faster than a Chebyshev expansion. A numerical example given below is the expansion of a Gaussian function

$$f(r) = \exp(-r^2), \quad 0 \leq r \leq 6. \tag{4.26}$$

This function is periodic in the sense that $f(-r) = f(r)$, and has an infinite number of derivatives. The Fourier expansion is based on the result of the integral $\int_0^\infty \exp(-x^2) \cos(b_n x) dx = [\sqrt{\pi}/2] \exp(-0.25 b_n^2)$, with $b_n = (\pi/12)(2n + 1)$, and $n = 0, 1, 2, \ldots$ so chosen that $\cos(b_n 6) = 0$. The Chebyshev expansion is performed with 21 Chebyshev functions, using a mesh grid as the zeros of the Chebyshev polynomial $T_{21}(x)$ in the interval $[-1, 1]$. The upper limit 6 of the radial interval is chosen large enough such that $\exp(-6^2) = 2.3 \times 10^{-16}$ is smaller than the numerical accuracy of MATLAB. The comparison of the two expansions is illustrated in Fig. 4.1. It shows that in this case the Fourier expansion converges significantly faster than the Chebyshev expansion.

4.4.1 Assignment

7.1: Consider the function given in Eq. (4.26). Expand this function into Lagrange polynomials based on a set of Lobatto support points, and examine the rate of convergence of the expansion coefficients and the error of the expansion in terms of the order N of these polynomials. Compare the rate of convergence with the results displayed in Fig. 4.1.

Fig. 4.1 Comparison of the
Fourier and Chebyshev
expansion coefficients for
$f(r) = \exp(-r^2)$,
$0 \le r \le 6$. The text
describes the discretization
of either expansion

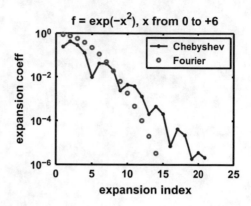

7.2: Repeat Assignment 7.1, but this time for the expansion into Lagrange polynomials use the following Chebyshev support points

$$x_j = \cos(j\pi/N), \quad j = 0, 1, \ldots, N.$$

References

1. L.N. Trefethen, *Spectral Methods in MATLAB* (SIAM, Philadelphia, 2000)
2. B.D. Shizgal, *Spectral Methods in Chemistry and Physics. Applications to Kinetic Theory and Quantum Mechanics* (Springer, Dordrecht, 2015)
3. A. Deloff, Ann. Phys. (NY) **322**, 1373–1419 (2007)
4. S. Širca, M. Horvat, *Computational Methods for Physicists*. Graduate Texts in Physics (Springer, Berlin, 2012)
5. Y.L. Luke, *Mathematical Functions and Their Approximations* (Academic, New York, 1975)

Chapter 5
Chebyshev Polynomials as Basis Functions

Abstract In the present chapter some of the important properties of Chebyshev polynomials are described, including their recursion relations, their analytic expressions in terms of the powers of the variable x, where $-1 \leq x \leq 1$, and the mesh points required for the Gauss–Chebyshev integration expression described in Chap. 3. We also point out the advantage of the expansion into this set of functions, as their truncation error is spread uniformly across the $[-1, 1]$ interval with the smallest error for functions that do not have strong singularities. The convergence of the expansion of functions in terms of Chebyshev polynomials will be illustrated, as well as the accuracy of the calculation of integrals and of derivatives. A novel "hybrid" method for calculating derivatives of higher order will also be described.

5.1 The Objective and Motivation

According to the polynomial expansion theorems described in Chap. 4, Chebyshev polynomials provide the expansion with the smallest error, which at the same time is uniform in the interval $[-1 \leq x \leq 1]$. In the present chapter these properties will be illustrated by means of examples. Furthermore, since integrals over Chebyshev functions can themselves be expressed in terms of Chebyshev functions, integrals over the function being expanded can be carried out easily and accurately. Hence Chebyshev polynomial expansions are also very useful for solving integral equations.

5.2 Some Properties of Chebyshev Polynomials

The variable x of the Chebyshev Polynomials $T_v(x)$, $v = 0, 1, 2, \ldots$, is contained in the interval $x \in [-1, 1]$, and is related to an angle θ according to

$$T_n = \cos(n\,\theta); \quad 0 \leq \theta \leq \pi, \tag{5.1}$$

© Springer Nature Switzerland AG 2018
G. Rawitscher et al., *An Introductory Guide to Computational Methods for the Solution of Physics Problems*,
https://doi.org/10.1007/978-3-319-42703-4_5

Fig. 5.1 Plots of the
polynomials $T_v(x)$. The
symbols denote the values at
the 14 mesh points ξ_k, which
are the zero's of T_{14}. They
are shown in order to
demonstrate that these points
are not equispaced

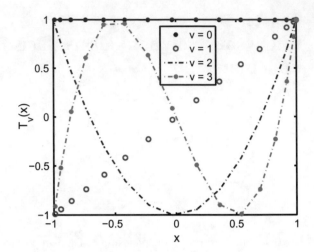

Fig. 5.1 Plots of the polynomials $T_v(x)$. The symbols denote the values at the 14 mesh points ξ_k, which are the zero's of T_{14}. They are shown in order to demonstrate that these points are not equispaced

where $x = \cos\theta$. This equation shows that the $x's$ are projections on the x-axis of the tip of a radius vector of unit length that describes a semi-circle in the x–y plane as θ goes from 0 to π. In terms of the x-variable the T_n's are given by

$$
\begin{aligned}
T_0 &= 1, \\
T_1 &= x, \\
T_2 &= 2x^2 - 1, \\
T_{n+1} &= 2xT_n - T_{n-1}.
\end{aligned}
\tag{5.2}
$$

It is clear from Eq. (5.1) that $-1 \le T_n(x) \le 1$, and that the larger the index n, the more zeros these polynomials have, as is shown in Fig. 5.1.

$$
\mathscr{I}_{n,m} = \int_{-1}^{+1} T_n(x)\, T_m(x)\, (1-x^2)^{-1/2}\, dx = \int_0^\pi \cos(n\theta)\,\cos(m\theta)\, d\theta,
\tag{5.3}
$$

with

$$
\mathscr{I}_{n,m} = \begin{cases}
0, & \text{for } m \ne n \\
\pi/2, & \text{for } n = m \ne 0 . \\
\pi, & \text{for } n = m = 0
\end{cases}
\tag{5.4}
$$

These polynomials also obey a discrete orthogonality relation

$$
\frac{\pi}{N} \sum_{k=1}^{N} T_n(\xi_k)\, T_m(\xi_k) = \frac{\pi}{2}\delta_{n\,m}(1 + \delta_{0\,n}) \quad n < N,\ m < N,
\tag{5.5}
$$

where the ξ_k are the zero's of T_N, given by

$$\xi_k = \cos\left[\frac{\pi}{N}(k - 1/2)\right], \quad k = 1, 2, \ldots, N. \tag{5.6}$$

The sum in Eq. (5.5) can also be expressed as

$$\sum_{k=1}^{N} T_n(\xi_k) \, T_m(\xi_k) = \begin{cases} 0, & \text{if } n, m < N, \text{ and } n \neq m \\ N/2, & \text{if } n = m, \text{ and } 0 < m < n, \\ N, & \text{if } n = m = 0 \end{cases} \tag{5.7}$$

according to Ref. [1] Chapter 11.7. The set defined in Eq. (5.6) (also called mesh-points) are the ones used both by Deloff [2] and in our own work, because they do not attain the values ± 1, thus avoiding possible singularities in the functions being calculated at these points. Even though the support points (5.6) do not include the end points, the expansion of a function (5.13) presented next in this section is valid for all values of x, including the end points, and integrals also include the end points. By contrast to Eq. (5.6), the support points used by Trefethen [3] are

$$x_k = \cos\left[\frac{\pi}{N}k\right], \quad k = 0, 1, 2, \ldots, N. \tag{5.8}$$

They do include the points ± 1, and are especially useful to construct differentiation matrices for functions that are not periodic, and also to calculate fast Fourier transforms (FFT), as described in Chapter 3 of Ref. [3]. A short description of Chebyshev polynomials and some of their properties is given in section 2.4.9 in Ref. [4], and a more extensive description can be found in the book by Luke [1].

The Chebyshev polynomials can be calculated as a function of x by either using Eq. (5.1), with $\theta = \arccos(x)$, or by means of the recursion relations

$$T_n(x) = 2x \, T_{n-1}(x) - T_{n-2}(x), \quad n \geq 2. \tag{5.9}$$

As can be seen from Fig. 5.1, at the end points their values are

$$T_n(1) = 1, \text{ and } T_n(-1) = (-)^n \text{ for all } n. \tag{5.10}$$

Note that in contrast to the zeros of Legendre or Laguerre polynomials, the support points (5.6) can be obtained analytically. Other useful properties of Chebyshev polynomials exist; for example,

$$2T_m(x)T_n(x) = T_{m+n}(x) + T_{|m-n|}(x), \tag{5.11}$$

contained in Chapter XI of Ref. [1], and

$$\int_0^1 T_{2n+1}(x) \sin(\alpha x) \frac{dx}{(1-x^2)^{1/2}} = (-1)^n \frac{\pi}{2} J_{2n+1}(\alpha) \quad (\alpha > 0),$$

$$\int_0^1 T_{2n}(x) \cos(\alpha x) \frac{dx}{(1-x^2)^{1/2}} = (-1)^n \frac{\pi}{2} J_{2n}(\alpha) \quad (\alpha > 0), \tag{5.12}$$

where the $J's$ are Bessel Functions, as described in Eq. (7.355) of Ref. [5].

Given a function $f(x)$, if one desires to expand it in terms of $N + 1$ Chebyshev polynomials, $T_n, n = 0, 1, 2, \ldots, N$,

$$f^{(N)}(x) = \sum_{n=1}^{N+1} a_n T_{n-1}(x), \, -1 \le x \le 1, \tag{5.13}$$

one obtains functions $f^{(N)}$ that are polynomials of order N whose values are equal to the values of f at all the $N + 1$ support points (which are the zeros of T_{N+1}). In the matrix notation given previously in Chap. 3, Eq. (3.19),

$$(f) = C \, (a) \tag{5.14}$$

and

$$(a) = C^{-1}(f). \tag{5.15}$$

The objects in parenthesis are column vectors of length $N + 1$, and the quantities C and C^{-1} are matrices. Because of the discrete orthogonality between Chebyshev polynomials, the $(N + 1) \times (N + 1)$ matrices C and C^{-1} are known and given in terms of Chebyshev Polynomials evaluated at the support points ξ_k, $k = 1, 2, \ldots, N + 1$. Hence, in order to obtain the expansion coefficients a_n, $n = 1, 2, \ldots, N + 1$, all what is needed are the discrete values of the function f at the zeros of T_{N+1}, given by Eq. (5.6).

Values of C and C^{-1} can be found in Ref. [6], are available in the list of appended MATLAB programs, and are routinely made use of in our numerical calculations. Interpolation of a function known only at the discrete ξ_k points to all values of x is obtained by means of Eq. (5.15), that can be written as

$$f^{(N)}(x) = \sum_{i=1}^{N+1} \sum_{k=1}^{N+1} T_{i-1}(x)(C^{-1})_{i,k} f(\xi_k). \tag{5.16}$$

Many additional properties are given in text books [7].

5.3   Integrals over Functions

Another important matrix relation applies to obtaining the integrals

$$F_L(t) = \int_{-1}^{t} f(x)dx \quad \text{and} \quad F_R(t) = \int_{t}^{1} f(x)dx. \tag{5.17}$$

Here the subscripts R and L stand for "left" and "right" respectively. From the expansion coefficients (a) of f, one can obtain the expansion coefficients $b^{(L)}$ or $b^{(R)}$ of the functions $F_L(t)$ or $F_R(t)$, respectively:

$$F_L^{(N)}(t) = \sum_{k=1}^{N+1} b_k^{(L)} T_{k-1}(t) \quad \text{and} \quad F_R^{(N)}(t) = \sum_{k=1}^{N+1} b_k^{(R)} T_{k-1}(t), \tag{5.18}$$

from the matrix relations

$$(b^{(L)}) = S_L^{(N)}(a) \quad \text{and} \quad (b^{(R)}) = S_R^{(N)}(a). \tag{5.19}$$

Again, these matrices S_L and S_R are standard for Chebyshev expansions, are available in our programs, and depend only on the value of the number of Chebyshev expansion polynomials, and their associated support points $\xi_k, k = 1, 2, \ldots, N+1$ in the open domain $(-1, 1)$.

If the definite integral $\mathfrak{I}$,

$$\mathfrak{I} = \int_{-1}^{1} f(x)dx, \tag{5.20}$$

is required, then, by making use of the expansion (5.18) for $F_L^{(N)}(t)$ for $t = 1$, and remembering that $T_k(1) = 1$ for all k, one finds

$$\mathfrak{I} \simeq \mathfrak{I}^C = \sum_{k=1}^{N+1} b_k^{(L)}. \tag{5.21}$$

This result should be identical to the Gauss–Chebyshev integral expression

$$\mathfrak{I} \simeq \mathfrak{I}^{GC} = \sum_{k=1}^{N+1} c_k f(\xi_k), \tag{5.22}$$

where

$$c_k = \int_{-1}^{1} \mathscr{L}_k(x)dx, \quad k = 1, 2, \ldots, N. \tag{5.23}$$

Expression (5.23) comes from the expansion of the polynomial $P_{N-1}(x)$,

$$f^{(N)}(x) = P_{N-1}(x) = \sum_{k=1}^{N+1} \mathscr{L}_k(x) f(\xi_k). \tag{5.24}$$

This quantity provides a polynomial approximation to the function $f(x)$. It is also called "interpolating polynomial" because it provides a value for $f^{(N)}$ for any point $x \in [-1, 1]$, and hence $\int_{-1}^{+1} f^{(N)}(x)dx = \sum_{k=1}^{N+1} [\int_{-1}^{1} \mathscr{L}_k(x)dx] f(\xi_k)$ from which results (5.22) and (5.23) follow. The $\mathscr{L}_k$ are Lagrange polynomials of order $N - 1$, and they vanish at all support points ξ_j with $j \neq k$. Furthermore, the $\mathscr{L}_k$ are orthogonal polynomials, with $\int_{-1}^{+1} \mathscr{L}_k(x)\mathscr{L}_j(x)dx = \delta_{k\,j}$, as can be seen in Ref. [8], Eq. (3). The above results are general expressions valid for orthogonal expansion polynomials. For each type of polynomial there is an associated particular set of support points that are the zeros of the orthogonal polynomial of one order higher than the ones used in the expansion of the interpolating polynomial.

For example, for Laguerre polynomials one has

$$\int_0^\infty e^{-x} f(x)dx \simeq \sum_{k=1}^{N+1} c_k f(\xi_k), \tag{5.25}$$

where

$$c_k = \int_0^\infty e^{-x} \mathscr{L}_k(x)dx, \quad k = 1, 2, \ldots, N + 1. \tag{5.26}$$

In MATLAB there are functions that give the weights and support points for Legendre functions $GLTable(nnode)$ or $GLNode(nnode)$, and $GaussLagQuad$ for Laguerre polynomials.

5.3.1   Assignments

5.1(a): Start from an equispaced discrete set of angles θ contained between 0 and π. Use the symbols '*' or something similar for your discrete points in the graphs below.

5.2(a): Calculate the corresponding set of x-values, and plot x versus θ.

5.3(a): Calculate the values of $T_n(x)$ for $n = 0, 1, 2$, and 3, and plot them as a function of x.

5.4(a): Calculate the values of $T_n(x)$ for $n = 0, 1, 2$, and 3, and plot them as a function of θ.

5.5(a): Check whether the forward recursion relation in Eq. (5.9) can be trusted, by comparing the result with Eq. (5.1).

5.6(a): Check which of the two integration methods, given by Eq. (5.21) or (5.21), produce a result with a higher accuracy. For this purpose choose any test function whose integral is known analytically.

Fig. 5.2 The Chebyshev
expansion coefficients a_n as
a function of the index n,
defined in Eq. (5.13), for the
two functions f_1 and f_2

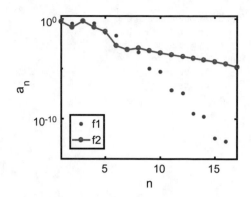

5.4 Examples of Chebyshev Expansions

By means of examples we will examine the rapidity and quality of the convergence
with N. In agreement with the theorems presented in Chap. 4 we will find that the
rapidity of the convergence depends on the "smoothness" of f, and in particular how
many singularities this function or its derivatives can have. The same procedure can
also be applied to expansions in terms of other basis functions [9]. In Chapter 4 of
Ref. [4] the convergence properties of expansions in terms of Hermite, Laguerre and
Chebyshev polynomials of various functions are illustrated by means of numerical
examples. However in this book the expansion coefficients are obtained by the calculation of overlap integrals, rather than by the matrix method of Curtis–Clenshaw
method [10], described in Eqs. (3.21) and (5.15). The support points are also different
from the ones we use, which for the Chebyshev expansions are given by Eq. (5.6).
The important question is: how fast will the error $|f(x) - f^{(N)}(x)|$ decrease with N?
According to Theorem 4 in section 4.4, the more singular the function or its derivative, the more slowly the expansion converges. Numerical examples for expansions
into Chebyshev polynomials confirm this prediction, as will be shown below.

These examples obtain the Chebyshev expansion coefficients for the functions

$$f_1(r) = r^{1/2} \sin(r), \tag{5.27}$$
$$f_2(r) = r \sin(r). \tag{5.28}$$

The results are displayed in Fig. 5.2. Since $f_1(r)$ is not an analytic function of r
(the first derivative has a singularity for $r = 0$) the expansion coefficients decrease
with the index ν much more slowly than for the analytic function $f_2(r)$. This result
was obtained by using the MATLAB functions $[C, CM1, xz] = C\_CM1(N)$ and
$r = mapxtor(b1, b2, xz)$.

An expansion into a Fourier series of the function $r \sin(r)$ is also carried out
for comparison with the expansion into Chebyshev polynomials. Fourier expansions
are described in Eqs. (5.38)–(5.45). Reference [4] devotes sections 4.6 to Fourier
expansions and Fourier transforms in much more detail than what is presented here.

Fig. 5.3 Comparison of the
Chebyshev and Fourier
expansion coefficients of the
function $x \sin(x)$, with
$0 \le x \le \pi$. In the text the
Fourier expansion
coefficients are denoted as c_n

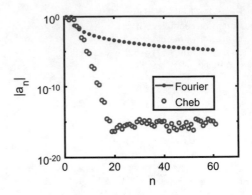

In our present example the Fourier expansion functions in the interval $[0, \pi]$ are $\sqrt{2/\pi} \sin(kr)$, $k = 1, 2, \ldots, k_{max}$. One finds that all c_n (the expansion coefficients in Eq. (5.44)) vanish for n odd, with the exception for $n = 1$, for which

$$c_1 = \frac{\pi^2}{4} \sqrt{\frac{2}{\pi}}. \tag{5.29}$$

For n even, the corresponding result for a_n is

$$c_n = \sqrt{\frac{2}{\pi}} \left[\frac{1}{(1+n)^2} - \frac{1}{(1-n)^2} \right], \quad n = 2, 4, 6, \ldots \tag{5.30}$$

For $n \gg 1$, c_n will approach 0 like $-4\sqrt{\frac{2}{\pi}}(1/n)^3$, i.e., quite slowly. By comparison with Fig. 5.2, or directly in Fig. 5.3, one sees that the Fourier expansion coefficients decrease with the index n much more slowly than the Chebyshev expansion coefficients, and hence the truncation errors of the expansion also decrease more slowly with the expansion index than for the Chebyshev expansion case.

5.4.1 An Estimate of the Error of a Chebyshev Expansion

That error is the error of truncating the expansion at a certain value of $N + 1$ terms. Theorems on the error are given in Chap. 4. Here we illustrate the rule of thumb, namely, that the truncation error is proportional to the absolute value of the next to the last expansion coefficient a_{N+2}, once the expansion begins to converge very rapidly.

In order to demonstrate this property, we reconsider the expansion of the function

$$f(x) = e^x, \quad -1 \le x \le 1. \tag{5.31}$$

Fig. 5.4 The Chebyshev
expansion coefficients a_k of
$\exp(x) = \sum_{k=1}^{N} a_n T_{n-1}(x)$
in the interval $[-1, 1]$, with
$N = 16$

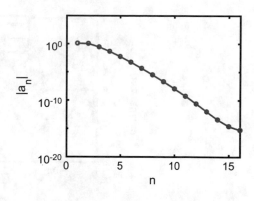

Fig. 5.5 The polynomial
approximant $f^{(2)}$ (dashed
blue curve) to the function
$f(x) = \exp(x)$ (green solid
curve) is compared to $f(x)$.
The slight deviation, given
by the error ε_2, is barely
visible in the figure. The
three support points are also
shown

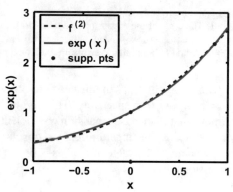

The Chebyshev expansion coefficients a_i are illustrated in Fig. 5.4, and the error
ε_N of the expansion is defined in

$$f(x) = \sum_{n=1}^{N+1} a_n T_{n-1}(x) + \varepsilon_N. \tag{5.32}$$

The approximation $f^{(N)} = \sum_{n=1}^{N+1} a_n T_{n-1}(x)$, given by Eq. (5.13) is a polynomial
of order N. This polynomial equals the function $f(x)$ at the $N + 1$ support points ξ_i,
which are the zeros of T_{N+1}. For example, for $N = 2$ the approximant polynomial
of order 2 is given by $f^{(2)}(x) = a_1 T_0(x) + a_2 T_1(x) + a_3 T_2(x)$, the 3 support points
are given by the zeros of T_3, and the error ε_2 is close to a_4. For the case that $f(x) =
\exp(x)$, $-1 \le x \le 1$, the function $f^{(2)}$ is compared with f in Fig. 5.5, that also
shows the position of the support points. The error of the expansion is seen in that
the dashed curve slightly disagrees with the solid curve.

In order to show that this error is proportional to the last expansion coefficient
a_{N+2}, the ratio ε_N / a_{N+2}, denoted as "normalized error", is displayed in Fig. 5.6 for
three values of N, for the case $f(x) = \exp(x)$.

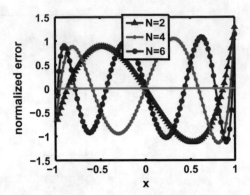

Fig. 5.6 The error ε_N/a_{N+2} of a Chebyshev expansion of the function $f(x) = exp(x)$, for $-1 \leq x \leq 1$. The error of the expansion is defined in Eq. (5.32), which includes polynomials up to T_N, and a_{N+2} is the last expansion coefficient for an expansion (5.13) with upper limit $N + 2$. For $N = 2, 4$, and 6 these relative errors are indicated by the blue, green, and red curves

The values of a_{N+2} in this case are $4.4 \times 10^{-2}, 5.4 \times 10^{-4}$, and 3.2×10^{-6} for $N = 2, 4$, and 6, respectively. This figure shows not only that the error is indeed proportional to the last expansion coefficient, but also that the error is uniform in the variable x, i.e., the upper limit of the absolute value of the normalized error is nearly independent of the value of x. This is one of the interesting properties of Chebyshev expansions.

5.4.2 Assignments

5.1(b): Consider the function

$$f(r) = r^{1/2} \sin(r), \quad 0 \leq r \leq \pi. \tag{5.33}$$

Define a new variable x that goes from -1 to $+1$ and relate it to the variable r by means of the linear transformation

$$r = ax + b$$

and find the values of a and b. Call the new function $\bar{f}(x) = f(r)$.

5.2(b): In preparation for expanding this function into Chebyshev polynomials choose a value of $N = 8$, and find the zeros ξ_i of $T_N(x)$ with $k = 1, 2, \ldots, N$. Use the expressions given in class

$$\xi_k = \cos\left[\frac{\pi}{N}(k - 1/2)\right], \quad k = 1, 2, \ldots, N, \tag{5.34}$$

and check that $T_N(\xi_k) = 0$ for all values of k.

5.3(b): Use the MATLAB functions $[C, CM1, xz] = C\_CM1(N)$ given in Appendix A, and check whether the output vector xz agrees with the vector of the ξ_k values obtained in 5.2(b). Note, xz is a column vector.

5.4(b): For each of the ξ_i values, obtain the corresponding r_k values. For that purpose use the function $r2 = mapxtor(b1, b2, xz)$ with $b1 = 0$ and $b2 = \pi$, where xz was obtained in part 5.3(b) to obtain $r2$ and check whether the values of r_k and $r2$ agree. Note that $r2$ should be a column vector.

5.5(b): Obtain the column vector $(F) = f(r2)$ and calculate the column vector (A) by means of the matrix×column vector MATLAB operation

$$(A) = CM1 \times (F). \tag{5.35}$$

The A-vector contains the coefficients a_n of the expansion (5.28). Check how fast the coefficients a_n decrease with n, with $n = 1, 2, \ldots, N + 1$.

5.6: Repeat parts 3, 4, and 5 for $N = 16$.

5.7: Define a new function

$$g(r) = r\, \sin(r), \quad 0 \le r \le \pi, \tag{5.36}$$

and repeat parts 3, 4, and 5 with $N = 16$. Check that the new expansion coefficients a_n decrease much faster with n than for the expansion of $f(r)$.

5.5 Derivatives of a Function

It is assumed for the discussions below that the expansion of the function into Chebyshev polynomials is known. Three methods to calculate the derivative of a given function are presented below: a hybrid, a Fourier and a Chebyshev method. The hybrid method consists in obtaining the values of the function by Chebyshev interpolation at three (or four) additional points surrounding each support point ξ_n. These points are located on an equispaced mesh at $\xi_n + h, \xi_n + 2h, \ \xi_n - h, \xi_n - 2h$, and the value of the various derivatives at ξ_n are obtained by standard finite difference formulas. The reason for this procedure is to avoid the large increase in value of the derivatives of the Chebyshev polynomials used in the Eq. (3.35) or (5.49). For example, to obtain the second order derivative, three points are used, $r_{n1} = \xi_n - h$, $r_{n2} = \xi_n$, and $r_{n3} = \xi_n + h$, and the values of the function at these points are denoted as f_{no}, f_{n1}, and f_{n2}, respectively. From these values, the finite difference expression is used for the second order derivative

$$d^2v/dr^2|_{\xi_n} = (f_{no} - 2f_{n1} + f_{n2})/h^2 + O(h^3), \quad n = 2, 3, \ldots, N. \tag{5.37}$$

These expressions can be found in Ref. [11], Table 25.2, or section 3.9 of Ref.
[4]. The method is called "hybrid" because it combines a spectral expansion method
with a finite difference method. Since the Chebyshev interpolation of the function
to equidistant points surrounding the Chebyshev support points can be done very
accurately, the error expected depends on the value of h.

The Fourier method consists in obtaining the coefficients of a Fourier expansion of
the given function in terms of the coefficients of the Chebyshev expansion. Since the
derivatives of the Fourier functions can be calculated analytically, the derivatives of
the given function can be obtained. The Chebyshev method consists in implementing
analytically the derivatives of the Chebyshev polynomials.

5.5.1 Connecting Chebyshev Space to Fourier Space

The expansion of a function into Chebyshev polynomials has very nice properties.
But taking the derivative with respect to the variable x may lead to inaccuracies
if the derivatives of the Chebyshev functions are involved, and if the Chebyshev
expansion coefficients a_v do not decrease with the index v fast enough. The reason is
that $|d^2 T_v(x)/dx^2|$, in Eq. (5.49), increases as v^4, as will be shown in Chap. 8 Fig. 8.1.
Hence the product $a_v d^2 T_v(x)/dx^2$ may not converge fast enough, and may introduce
truncation errors. In this case, migration to a Fourier space may be more suitable,
since the derivatives of the Fourier expansion functions $\Phi_v(x)$ increase with the index
v more slowly, like v^2 for the second derivative.

5.5.1.1 Fourier Expansion Obtained from a Chebyshev Expansion

In the present description the Fourier basis functions are cosine functions Φ_{n_f},
defined in the radial interval $[r_{min}, r_{max}]$ as

$$\Phi_{n_f}(r) = \cos\left[\frac{(r - r_{max})}{(r_{max} - r_{min})} n_f \pi\right], n_f = 0, 1, 2, \ldots, N_F, \tag{5.38}$$

which in the x-space $[-1, 1]$ becomes

$$\Phi_{n_f}(x) = \cos[(x - 1) n_f \pi/2], \tag{5.39}$$

where the r to x linear mapping is given by

$$2r = (r_{max} + r_{min}) + (r_{max} - r_{min}) x. \tag{5.40}$$

The corresponding Fourier sine functions are denoted as

$$\bar{\Phi}_{n_f}(x) = \sin[(x - 1) n_f \pi/2], \quad n_f = 0, 1, 2, \ldots, N_F. \tag{5.41}$$

The orthonormality of the $\Phi_{n_f}(x)$, defined by

$$\left(\Phi_{n_1} | \Phi_{n_2}\right) = \int_{-1}^{1} \Phi_{n_1}(x)\, \Phi_{n_2}(x) dx$$

is as follows

$$\left(\Phi_{n_1} | \Phi_{n_2}\right) = \begin{cases} 2, & \text{if } n_1 = n_2 = 0 \\ 1, & \text{if } n_1 = n_2 \neq 0 \ . \\ 0, & \text{if } n_1 \neq n_2 \end{cases} \tag{5.42}$$

Using 41 Chebyshev support points the numerical results for the $\left(\Phi_{n_1} | \Phi_{n_2}\right)$ orthogonality integrals, described in Eq. (5.42), are obtained with machine accuracy.

If an expansion of a function u in terms of Chebyshev polynomials T_{n_t} is given by

$$u(r) = \sum_{n_t=1}^{N_T+1} a_{n_t} T_{n_t-1}(x), \tag{5.43}$$

and if the expansion coefficients a_{n_t} are known, then the purpose of this discussion is to obtain the corresponding expansion in terms of Fourier functions

$$u(r) = \sum_{n_t=1}^{N_F+1} c_{n_f} \Phi_{n_f-1}(x). \tag{5.44}$$

For this purpose the overlap integrals of Fourier functions with Chebyshev functions T_{n_t}

$$Ov(n_f, n_t) = \int_{-1}^{1} \Phi_{n_f}(x) T_{n_t}(x)\, dx \tag{5.45}$$

are required. An illustration of the result is shown in Fig. 5.7. In the numerical MATLAB program that performs the integrals, $T_{n_t}(x)$ is given by $\cos[n_t \cdot a \cos(x)]$, and because the product $\Phi_{n_f}(x) T_{n_t}(x)$ is strongly oscillatory the number of Chebyshev support points used for the calculation of the overlap matrix $Ov(n_f, n_t)$ elements, depicted in Fig. 5.7, was 600. For $n_f > 360$ the results become unreliable. The figure also shows that when both n_f and n_t are large but nearly equal to each other (like $n_f \simeq n_t \simeq 50$), the overlap integral reaches its maximum value. The computing time in MATLAB to generate the Ov matrix displayed in Fig. 5.7 is between 1 and 2 s using a Windows 10 desktop computer, with an Intel i7 processor, a 2.80 GHz CPU and 8 GB of RAM.

The procedure for obtaining the Fourier expansion coefficients of the function u in terms of the Chebyshev expansion coefficients is as follows. By equating this expansion (5.43) to the Fourier expansion (5.44) the values of the Fourier expansion coefficients c_{n_f} are obtained in terms of Chebyshev expansion coefficients by

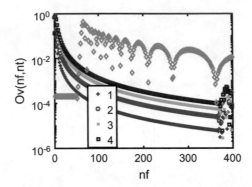

Fig. 5.7 The magnitude of the overlap integrals $\int_{-1}^{1} \Phi_{nf}(x) \cdot T_{n_t}(x)dx$ as a function of n_f for a fixed value of n_t. The lowest curve corresponds to $n_t = 1$, the next lowest to $n_t = 2, 3$, and 4. The highest curve (with lozenge markers) corresponds to $n_t = 100$. The total number of Chebyshev support points used to perform the overlap integral was 600. Many of the overlaps are zero, but they are suppressed in the graphs

integrating both expansions over the functions $\Phi_{n_f}(x)$, $n_f = 0, 1, 2, \ldots, N_F$. By making use of the orthogonality properties of the Fourier functions one finds the linear relationship

$$c_{n_f} = \sum_{n_t=1}^{N_T+1} a_{n_t} Ov(n_f, n_t). \tag{5.46}$$

The derivatives of the function u then emerge as $du/dr = 2(du/dx)/(r_{\max} - r_{\min})$ and $d^2u/dr^2 = 4(d^2u/dx^2)/(r_{\max} - r_{\min})^2$ with

$$du/dx = - \sum_{n_f=1}^{N_F+1} (n_{f-1} \pi/2)c_{n_f-1}\bar{\Phi}_{n_f-1}(x), \tag{5.47}$$

and

$$d^2u/dx^2 = - \sum_{n_f=1}^{N_F+1} (n_{f-1} \pi/2)^2 c_{n_f-1}\Phi_{n_f-1}(x). \tag{5.48}$$

The advantage of these expressions is that the coefficients $(n_{f-1} \pi/2)^2 c_{n_f-1}$ increase with n_f much more slowly than had the expansion of d^2u/dx^2 been done in terms of $d^2 T_{n_t}(x)/dx^2$, since the latter increase like $(n_t)^4$. A disadvantage of the Fourier expansions is that they converge much more slowly than the Chebyshev expansions. An example is given in the subsection below.

5.5.2 Numerical Examples for Calculating Derivatives

The derivatives in Chebyshev space are obtained from

$$d^k u / dr^k = \left[\frac{2}{(r_{max} - r_{min})} \right]^k \sum_{n_t=1}^{N_T+1} a_{n_t} d^k T_{n_t-1}(x)/dx^k, \quad k = 1, 2, \ldots, \quad (5.49)$$

while the derivatives in Fourier space are obtained by means of Eqs. (5.47) and (5.48). The number of Chebyshev and Fourier expansion coefficients is 61 and 101 respectively, and the value of h in Eq. (5.37) is 10^{-2} for all the calculations described below. A more extended discussion can be found in section 3.9.2 of Ref. [4]. In next subsection we present one example of a function without singularities.

5.5.2.1 A Function Without Singularities

A numerical test case is performed for the function

$$u = y/(1 + y)^2, \quad y = \exp[(r - R)/a]. \quad (5.50)$$

The first and second derivatives with respect to r can be obtained analytically from the above. The calculation is performed with

$$a = 0.5, \quad R = 4,$$

in the radial interval $[0, 10]$, with $n_f = 0, 1, 2, \ldots, 100$, and $n_t = 0, 1, 2, \ldots, 60$. The functions u, du/dr, d^2u/dr^2 and d^3u/dr^3 are illustrated in Fig. 5.8 at their Chebyshev support points.

Fig. 5.8 The test functions u, du/dr, d^2u/dr^2 and $d3u/dr^3$. The function u is defined in Eq. (5.50)

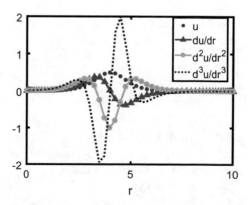

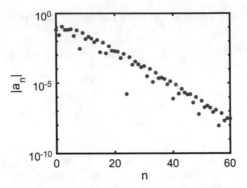

Fig. 5.9 The Chebyshev expansion coefficients of the test function u given by Eq. (5.50). Even though the function u has no singularity, the convergence of the expansion is "slow"

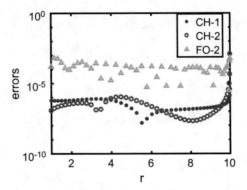

Fig. 5.10 The errors for du/dr and d^2u/dr^2 for the test function u, Eq. (5.50). The results obtained in Chebyshev space with 61 support points are labelled as CH-1 and CH-2, respectively. The errors for d^2u/dr^2 obtained in Fourier space with 101 Fourier functions are denoted as FO-2. The Fourier results are obtained with a piece of the Overlap Matrix illustrated in Fig. 5.7, of dimension $(NF + 1) \times (NT + 1) = 101 \times 61$

Table 5.1 Accuracies for the calculation of derivatives of the function u by various methods

Method	u	du/dr	d^2u/dr^2	d^3u/dr^3
Fourier	$10^{-7} \to 10^{-6}$	$10^{-6} \to 10^{-4}$	$10^{-4} \to 10^{-3}$	$10^{-2} \to 10^{-1}$
Hybrid	–	$10^{-7} \to 10^{-6}$	$10^{-6} \to 10^{-5}$	$10^{-4} \to 10^{-3}$
Chebyshev	$10^{-16} \to 10^{-15}$	$10^{-7} \to 10^{-6}$	$10^{-7} \to 10^{-6}$	$10^{-5} \to 10^{-4}$

The MATLAB program that calculates the derivatives is "test_deriv.m". The Chebyshev expansion coefficients a_{n_i} for $u(r)$ are illustrated in Fig. 5.9 and the resulting errors are illustrated in Fig. 5.10.

A comparison of the different types of errors is presented in Table 5.1. The results for the Chebyshev Method are more accurate than for the Hybrid Method, which in turn is more accurate than the Fourier Method.

In the next subsection we present one example of a function with singularity in the derivatives.

5.5.2.2 A Function with a Singularity in the Derivatives

The function for this test case is

$$v = r^{(1/2)} \cos(r), \text{ for } 0 \leq r \leq 10. \qquad (5.51)$$

This function, together with the first and second derivatives with respect to r is displayed in Fig. 5.11.

Because of the singularities of the first and higher derivatives near the origin, the radial interval is divided into two parts: $[0, \pi]$ and $[\pi, 10]$. The corresponding Chebyshev expansion coefficients a_n decrease with the index n relatively slowly in the first partition, but quite rapidly in the second partition, as shown in Fig. 5.12.

The question to be answered for this example is: which of the expansions for the second and third order derivatives converge faster: the Fourier, the Chebyshev or the Hybrid? The answer for both partitions is shown in Fig. 5.13 and is as follows.

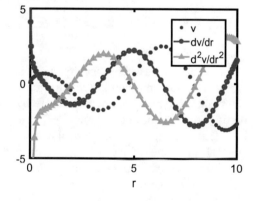

Fig. 5.11 Test function $v = r^{1/2} \cos(r)$, and the first and second derivatives dv/dr and d^2v/dr^2. Both derivatives become singular near the origin

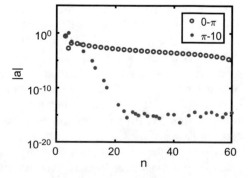

Fig. 5.12 The Chebyshev expansion coefficients for the function v in each of the radial domains $[0, \pi]$ and $[\pi, 10]$

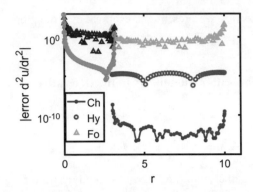

Fig. 5.13 Accuracy of the calculation of d^2v/dr^2 by three different methods, in each of the two radial partitions $[0, \pi]$ and $[3, 10]$. In the first partition the Chebyshev expansion of v converges much more slowly that in the second partition. The symbols in the legend Ch, Hy, and Fo refer to Chebyshev, Hybrid, and Fourier, respectively. In the first partition the Ch and Hy results are almost identical, hence indistinguishable in the figure. In the second partition the accuracy of the hybrid method is $\simeq 10^{-5}$, but it can be increased if h is made smaller

Table 5.2 Accuracies for the calculation of derivatives of the function v in partition $[0, \pi]$

Method	v	dv/dr	d^2v/dr^2	d^3v/dr^3
Fourier	$10^{-4} \to 10^{-3}$	$10^{-2} \to 10^{-1}$	$10 \to 10^1$	$1 \to 10^1$
Hybrid	–	$10^{-3} \to 10^{-2}$	$10^{-4} \to 10^{-2}$	$10^{-1} \to 10^2$
Chebyshev	10^{-15}	$10^{-4} \to 10^{-3}$	$10^{-4} \to 10^{-1}$	$1 \to 10^1$

The accuracy of the Chebyshev expansion is approximately the same as the accuracy for the Hybrid expansion in the partition $[0, \pi]$, while in the $[\pi, 10]$ partition the Chebyshev expansion is more accurate by several orders of magnitude. In either partition the accuracy of the Fourier expansion as indicated by the triangular symbols is unacceptably low. The numerical calculation is based on the MATLAB program "test_f1f2.m", the number of Chebyshev and Fourier expansion coefficients is 61 and 101 respectively, and the value of h in Eq. (5.37) is 10^{-2}.

The Fourier method is based on Eqs. (5.47) and (5.48). For the Chebyshev method the derivative matrix D, explained after Eq. (3.35), is given by CHder1 and CHder2 obtained from the MATLAB program "der2CHEB(xz).m" listed in the appendix. The hybrid method is based on Eq. (5.37).

The overall conclusion is as follows. The results for the $[0, \pi]$ partition, for which the Chebyshev expansion converges slowly, are summarized in Table 5.2. One can conclude that the Chebyshev method for obtaining derivatives is the most accurate one, and is comparable to the accuracy obtained by the hybrid method. The accuracy of the expansion of the function v in terms of Fourier functions is also acceptable. The results for the $[\pi, 10]$ partition, in which the Chebyshev expansion converges very fast, indicate that the Chebyshev expansion for the derivatives is the most accurate. The reason for the good accuracy of the hybrid method is due of the great accuracy of calculating the values of the function at arbitrary points by Chebyshev interpolation.

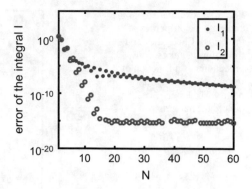

Fig. 5.14 Convergence properties of the Gauss–Chebyshev integration procedure as a function of the number of Chebyshev support points N in the interval $[0, \pi]$, for both the integrals $I_1 = \int_0^\pi r^{1/2} \sin(r) \, dr$ and $I_2 = \int_0^\pi r \sin(r) \, dr$. For 61 Chebyshev support points the error of I_1 and I_2 is 10^{-9} and 10^{-15}, respectively

5.6 The Chebyshev Error of the Integral of a Function

Approximations to integrals of a function can be obtained in terms of the Chebyshev expansion coefficients of the function, as is described in Eqs. (5.17)–(5.19) and (5.21). In these equations the matrices S_L and S_R play an important role. By summing the expansion coefficients of the indefinite integral using S_L one also obtains an approximation for the definite integral $I = \int_a^b f(r) dr$, as shown in Eq. (5.21). We will make much use of these matrices in Chap. 6. These properties are discussed extensively in Refs. [6, 12, 13]. The rule of thumb for the error of the integral is as follows. If the Chebyshev expansion of the function f converges slowly with N, then the error of the integral I also decreases slowly, and is of the same order of magnitude as the error of the Chebyshev expansion of the integrand f. This will now be demonstrated numerically. Chapter 3 of Ref. [4] contains detailed results for integrals and their errors and describes a different method based on mapping the function into different variables.

The error of the two definite integrals

$$I_1 = \int_0^\pi r^{1/2} \sin(r) \, dr \quad \text{and} \quad I_2 = \int_0^\pi r \sin(r) \, dr, \tag{5.52}$$

obtained by means of Eqs. (5.17), (5.18), is displayed in Fig. 5.14. The numerical value of these integrals is 2.43532116417 and π, respectively. The former is obtained by means of the MATLAB program $Q = integral(f, 0, pi,' AbsTol', 1e - 10)$.

A comparison of the error of the integrals I_1 and I_2 with the corresponding errors of the finite difference Simpson method is displayed in Fig. 5.15. The figure shows that for the same number of mesh points, the accuracy of the Simpson method is less than that of the Chebyshev method. This difference is especially pronounced for the case of I_2. In the conclusions of this chapter, the properties of the expansions of functions and of their integrals in terms of Chebyshev polynomials are described. It is shown that the rate of convergence of the expansions depends on the properties of the function, and that the expansions of the integrals tend to converge faster than when performed by the Taylor expansion-based Simpson rule. Please note that the errors

Fig. 5.15 Comparison of the convergence properties of the Gauss–Chebyshev and the Simpson integration procedures as a function of the number of support points. The labels 1 or 2 denote the integrals $I_1 = \int_0^\pi \sin(r)\, r^{1/2}\, dr$ or $I_2 = \int_0^\pi \sin(r)\, r\, dr$, respectively

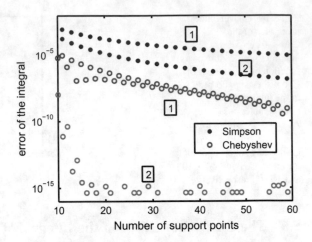

of the expansions in terms of Chebyshev polynomials are of a different nature than the errors that occur in the calculation of recursion relations, for example. The latter are due to the presence of a solution that increases with the number of iterations and overwhelm the decreasing solution. These errors are related to the finite number of significant figures that the computer carries, and occur sooner that the accumulation of round-off errors.

References

1. Y.L. Luke, *Mathematical Functions and Their Approximations* (Academic, London, 1975)
2. A. Deloff, Ann. Phys. (NY) **322**, 1373–1419 (2007)
3. L.N. Trefethen, *Spectral Methods in MATLAB* (SIAM, Philadelphia, 2000)
4. B.D. Shizgal, *Spectral Methods in Chemistry and Physics. Applications to Kinetic Theory and Quantum Mechanics* (Springer, Dordrecht, 2015)
5. L.S. Gradshteyn, I.M. Ryzhik, *Tables of Integrals, Series, and Products*, 4th edn. (Academic, New York, 1980)
6. R.A. Gonzales, J. Eisert, I. Koltracht, M. Neumann, G. Rawitscher, J. Comput. Phys. **134**, 134 (1997)
7. Y.L. Luke, *Mathematical Functions and Their Approximations* (Academic, New York, 1975); J.P. Boyd, *Chebyshev and Fourier Spectral Methods*, 2nd revised edn. (Dover Publications, Mineola, 2001)
8. D. Baye, Phys. Stat. Sol. **243**, 1095–1109 (2006); D. Baye, The Lagrange-mesh method. Phys. Rep. **565**, 1–107 (2015)
9. P.Y.P. Chen, Appl. Math. **7**, 927 (2016). https://doi.org/10.4236/am.2016.79083
10. C.C. Clenshaw, A.R. Curtis, Numer. Math. **2**, 197 (1960)
11. M. Abramowitz, I. Stegun (eds.), *Handbook of Mathematical Functions* (Dover, New York, 1972)
12. G.H. Rawitscher, I. Koltracht, Description of an efficient numerical spectral method for solving the Schröedinger equation. CiSE (Comput. Sci. Eng.) **7**, 58–66 (2005)
13. G.H. Rawitscher, Applications of a numerical spectral expansion method to problems in physics; a retrospective, *Operator Theory: Advances and Applications*, vol. 203 (Birkhauser Verlag, Basel, 2009), pp. 409–426

Chapter 6
The Integral Equation Corresponding to a Differential Equation

Abstract In this chapter we present an integral equation, whose solution is the same as that of a corresponding second order differential equation. We discuss the advantages of working with the integral equation, called Lippmann–Schwinger (L–S). We show how a numerical solution of such an equation can be obtained by expanding the wave function in terms of Chebyshev polynomials, and give an example for a simple one-dimensional Schrödinger equation. This method is denoted as S-IEM (for spectral integral equation method), and its accuracy is discussed. A case of a shape resonance is also presented, and the corresponding behavior of the wave functions for different incident energies is described.

6.1 Summary and Motivation

Given a second order differential equation, an equivalent integral equation exists called Lippmann–Schwinger (L–S), whose solution is the same as that of the differential equation. The implementation of the L–S method involves a Green's function, which is defined in coordinate space, as described below. The advantages of working with the L–S equation in coordinate space are numerous: (a) the Green's function involved has a singularity which is easier to implement numerically than the corresponding Green's function in momentum space. The latter is frequently described in quantum mechanics text books, such as in Eq. (7.13) and Eq. (8.12) in Ref. [1]; (b) the accumulation of round-off errors is substantially smaller than the accumulation of the round-off errors that occur for the solution of the differential equation with a finite difference method, as is shown in Fig. 6.1; (c) the derivative of the solution of the L–S equation can be obtained in terms of integrals whose accuracy is higher than the derivatives obtained by finite difference methods, as will be shown for the solutions based on Chebyshev expansions; (d) the effect of small potentials or other small perturbations can be included iteratively in a natural way. An example is the Born series, already known by physicists for a long time, and frequently used for quantum-mechanical calculations. A difficulty in solving the integral equation is that the matrices involved are generally non-sparse (in contrast to the matrices involved

© Springer Nature Switzerland AG 2018
G. Rawitscher et al., *An Introductory Guide to Computational Methods
for the Solution of Physics Problems*,
https://doi.org/10.1007/978-3-319-42703-4_6

Fig. 6.1 Comparison of the
error of the solution of a
spherical Bessel function
equation by a finite
difference method and a
spectral integral equation
method. Details of the
calculation are described in
the text, and also in Ref. [2]

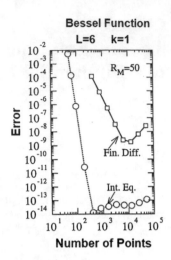

in finite difference methods) and hence present a higher numerical complexity than finite difference methods. This difficulty can be overcome by dividing the radial range into partitions, thus reducing the dimension of the matrices in each partition to a useful level [2].

In order to motivate the explanation of the spectral integral method (S-IEM) for solving a second order differential equation, two examples are given below that compare the S-IEM method with a finite difference Numerov method. The first example, shown in Fig. 6.1, consists of the solution $j_L(x)$ of the spherical Bessel differential equation, Eq. (10.1.1) in Ref. [3], as described in Ref. [2], Fig. 1. This solution was normalized by comparison with a known tabulated function at one particular point in the radial domain [0, 50].

The angular momentum number is $L = 6$, i.e., the potential V is given by $L(L + 1)/r^2$, and the wave number $k = 1$ is in units of inverse length. The singularity of V at the origin did not have to be treated especially in the S-IEM method, since the Chebyshev mesh points did not reach $r = 0$. However, for the finite difference method, the solution had to be started near the origin by expanding j_L in powers of r. Both calculations are done in FORTRAN with double precision. The curve marked as "Fin Diff" is obtained via the finite difference Numerov method, with an error of order h^6 in each three-point recurrence relation, where h is the distance between radial mesh points. This method is given by Eq. (25.5.21) in Ref. [3], where it is described as Milne's method. The curve labeled "Int Eq" was calculated with the S-IEM, without the imposition of an accuracy parameter tol. Instead, the radial domain was divided into partitions of equal length, each containing $N + 1 = 17$ Chebyshev support points. As the size of each partition was decreased manually, the number m of partitions in the radial domain [0, 50] increased correspondingly, and hence the total number of support points $m \times 17$, displayed on the x-axis, increased correspondingly. If an accuracy parameter had been given, then the size of each partition containing 17 mesh points would have been adjusted adaptively in each partition so as to provide the accuracy requested. This would have led to small partitions in the region where the function is strongly oscillatory, and large partitions in the region where the function is less oscillatory. This is demonstrated further in Fig. 3 of Ref. [4].

Figure 6.1 shows three important properties of the S-IEM method in comparison with the finite difference method: (a) the error decreases much more rapidly with the number of support points, which demonstrates the super-algebraic reduction of the truncation error with the size of the partition; (b) the accumulation of the round-off error is much slower than what is the case for the finite difference method, as can be seen from the slope of the curves beyond the minima; and (c) the maximum accuracy achieved before the round-off errors overwhelm the algorithm is much higher. This last remark is due to the confluence of two facts:

1. Since there are fewer mesh points for the S-IEM method than for finite difference methods for a given accuracy, the accumulation of round-off errors is less and hence the overall accuracy is higher;
2. The number of mesh points beyond the region where the accumulation of round-off errors begins to dominate over the accumulation of algorithm errors is also much smaller. Additional figures are shown in Ref. [2].

A second example is given by the accuracy of the phase shift φ of a wave function which is the solution of a Schrödinger equation in the case of a resonance phenomenon at a barrier, and is described in detail in Ref. [5], and also in the resonance section of the present chapter. The phase-shift of a wave, defined in Eq. (7.6) (where it is denoted as δ) is a quantity that describes the asymptotic behavior of the wave, and the error of φ is a measure of the accumulation of errors in the course of the calculation of the wave function. The accuracy of φ obtained with the S-IEM method as compared with the accuracy obtained with a finite difference method is taken from Ref. [4], and is illustrated in Fig. 6.2. The curves labelled "LD" and "NUM" are obtained with a logarithmic derivative method and the Numerov method, respectively, and "IEM" curves are obtained with two different versions of the S-IEM method. The figure shows that the accuracy of the S-IEM method is higher than the other methods by approximately six orders of magnitude. The reason for this difference in accuracy is due to the fact that in a barrier region, where the potential is larger than the energy, there are two solutions to the Schrödinger equation, one that increases exponentially with distance and the other that decreases exponentially. At the peak of the resonance energy the wave function decreases, but the numerical errors introduce an exponentially increasing component. These errors are substantially smaller for the S-IEM method, hence the better accuracy.

A comparison between the S-IEM and Numerov methods of the computing times is displayed in Tables 4.2 and 4.3. For the same accuracy of the result, the Numerov method can be several orders of magnitude slower, depending on the accuracy required.

Fig. 6.2 Comparison of the accuracies of two S-IEM methods and two finite difference methods for the calculation of phase-shifts in a resonance region. The energy of the incoming wave is given by k^2, the phase-shift φ is a measure of the asymptotic behaviour of a wave function, defined in Eq. (6.11). The potential is defined in Eq. (6.16), and is illustrated in Fig. 6.5

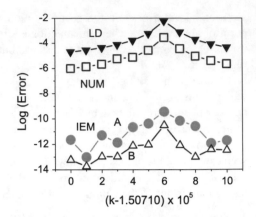

6.2 Introduction

The Schrödinger second order differential equation for a partial wave function $\psi(r)$ in one dimension has the form

$$\left(\frac{d^2}{dr^2} - L(L+1)/r^2 + k^2\right)\psi(r) = V(r)\psi(r), \quad 0 \le r \le \infty, \qquad (6.1)$$

that is usually solved by a finite difference method, such as Runge–Kutta [6, 7]. Here r is the radial distance, k^2 is the energy in units of inverse length squared (assumed given), k is the wave number, $V(r)$ (assumed given) is the local potential, also given in units of inverse length squared, and ψ is dimensionless. In order to convert the energy and the potential, previously given in energy units, to the units of inverse square length, one multiplies the former by the factor $2m/\hbar^2$, where $\hbar$ is Planck's constant and the reduced mass of the two interacting objects is m.

The solution $\psi(r)$ of Eq. (6.1) is also a solution of the integral equation, denoted as Lippmann–Schwinger [8] (L–S), of the form

$$\psi(r) = F(r) + \int_0^{R_{max}} \mathcal{G}_k(r, r') \, V(r') \, \psi(r') \, dr'. \qquad (6.2)$$

The Green's function $\mathcal{G}_k(r, r')$ and the boundary conditions for ψ will be described below. Here R_{max} is the value of a radial distance beyond which the potential V can be neglected compared to k^2 to within the desired accuracy. A demonstration of the equivalence of the solutions of Eqs. (6.1) and (6.2) is given in Ref. [2]. Most physicists usually prefer to solve the differential equation because of the simplicity of the numerical finite difference algorithm, and shy away from solving integral equations because the related matrices are non-sparse, hence computationally more demanding and more memory intensive. However $\mathcal{G}_k(r, r')$ in configuration space is much than in momentum space, since its singularity in configuration space is easier to handle computationally than its singularity in momentum space [9].

In configuration space, the Green's function $\mathscr{G}_k(r, r')$ for a second order differential equation in one dimension is given by [1]

$$\mathscr{G}_k(r, r') = -\frac{1}{k} F(r_<) G(r'_>), \tag{6.3}$$

with $r_<$ and $r_>$ being the lesser and larger values of r and r', respectively. Thus, the explicit form of Eq. (6.2) is

$$\psi(r) = F(r) - \frac{1}{k} F(r) \int_r^{R_{max}} G(r') \, V(r') \, \psi(r') \, dr'$$
$$- \frac{1}{k} G(r) \int_0^r F(r') \, V(r') \, \psi(r') \, dr', \tag{6.4}$$

where k is the wave number defined in Eq. (6.1) and where the functions F and G are the regular and irregular spherical Bessel functions, solutions of the same equation but subject to different boundary conditions

$$\left(d^2/dr^2 - L(L+1)/r^2 + k^2 \right) F(r) = 0 \tag{6.5}$$

and

$$\left(d^2/dr^2 - L(L+1)/r^2 + k^2 \right) G(r) = 0.$$

They are linearly independent of each other, and F approaches 0 as $r \to 0$, while $|G| > 0$ as $r \to 0$. If the angular momentum quantum number $L = 0$, then

$$F(r) = \sin(kr), \quad G(r) = \cos(kr). \tag{6.6}$$

It is left as a homework exercise to show that the solution of Eq. (6.4) also obeys Eq. (6.1), and further, that the derivative $\psi'(r) = d\psi(r)/dr$ is given by

$$\psi'(r) = F'(r) - \frac{1}{k} F'(r) \int_r^{R_{max}} G(r') \, V(r') \, \psi(r') \, dr'$$
$$- \frac{1}{k} G'(r) \int_0^r F(r') \, V(r') \, \psi(r') \, dr'. \tag{6.7}$$

In the derivation of Eq. (6.7) the terms due to the derivatives of the integrals cancel each other. This equation is very useful because it contains the derivatives of known functions F and G, and the rest is done by integrals that in the spectral method do not lose hardly any accuracy, as discussed in Chap. 5, and in Ref. [4]. One of the reasons is that the accumulation of round-off errors is less than in the case of finite difference methods, as illustrated in the introduction. Another reason is that a spectral expansion tends to be more accurate and controllable than in the case of finite difference methods. The solution of Eq. (6.4), obtained by expanding the wave function ψ in terms of Chebyshev polynomials, is denoted by S-IEM (for spectral integral equation method) and is described in detail as follows.

In practice, it is convenient to group the $-L(L+1)/r^2$ term together with the potential, in this case replacing V in Eq. (6.4) by V_L

$$V_L = V + L(L+1)/r^2 \,,$$

and using for F and G the expressions given by Eq. (6.6). It has been verified that in this case the singularity of $L(L+1)/r^2$ near the origin in V_L does not cause numerical difficulties for the S-IEM method based on Chebyshev expansions. The verification, performed in Ref. [2], consists in solving Eq. (6.5) by using V_L in Eq. (6.4). The accuracy of the result is illustrated in Fig. 1 of Ref. [2], and is reproduced in Fig. 6.1. As mentioned in the introduction, this figure shows the much larger accuracy obtained with the S-IEM method, compared with a Numerov method for the same number of support points, and that the accumulation of round-off errors is also much smaller. The S-IEM can be extended to the case of coupled radial equations, as shown in Ref. [10].

Asymptotically, when $r \geq R_{max}$ the solution of Eq. (6.4) is given by

$$\psi(r) = F(r) + T G(r), \tag{6.8}$$

with the dimensionless quantity T given by

$$T = -\frac{1}{k} \int_0^{R_{max}} F(r') \, V(r') \, \psi(r') \, dr'. \tag{6.9}$$

In the asymptotic region the wave function can be expressed in terms of a phase shift φ

$$\psi(r) = K \sin(kr - L\pi/2 + \varphi), \quad r \geq R_{max}, \tag{6.10}$$

where K is a normalization factor applied to the solution of Eq. (6.4). By comparing Eq. (6.8) with (6.10) together with a little of trigonometry, one finds that

$$\tan(\varphi) = T, \quad \text{and} \quad K = \sqrt{T^2 + 1}. \tag{6.11}$$

The quantity T plays an important role in scattering theory, since the phase shift is given by $\varphi = \arctan(T) + n\pi$. A drawback to obtaining a numerical result for the integral (6.9) in the case that $L \neq 0$, even for short ranged potentials V, is that the spherical Bessel functions for F and G have to be used in Eq. (6.4). Alternatively, if F and G are given by Eq. (6.6), then V in Eq. (6.9) has to be replaced by V_L, and the upper limit R_{max} in the integral (6.9) can become very large, unless the solution $\psi(r)$ is matched to a combination of spherical functions for $r \geq R_{max}$. Another way around that difficulty consists in using the Phase-Amplitude Method [11] described in Chap. 8.

The S-IEM method can also be adapted to find eigenvalues and eigenfunctions for bound states. In this case the Green's function consists of exponential functions, rather than sine and cosine functions, whose wave number is obtained from the negative energy, determined iteratively. This method is explained in Chap. 10. An example applied to the bound states of a He-He diatom is contained in Ref. [12].

6.3 Spectral Implementation of the S-IEM Method

The spectral solution of Eq. (6.4) has been described extensively in Refs. [4, 13], and is denoted as "S-IEM". The expansion into Chebyshev functions is given by Eq. (5.32), $\psi(x) = \sum_{k=1}^{N+1} a_k T_{k-1}(x)$. Here the variable $r \in [0, R_{max}]$ has already been replaced by the variable $x \in [-1, +1]$ by means of a linear transformation. Expansions into other sets of orthogonal polynomials, such as Laguerre or Hermite can also be considered, but the use of Chebyshev polynomials is especially suitable for carrying out the integrals involved in the integral equations, hence they are preferred in these situations.

A basic version proceeds as follows:

1. An upper integration limit R_{max} in Eq. (6.2) is chosen, such that beyond R_{max} the potential V is sufficiently small and can be neglected to within the accuracy desired; and the number $N + 1$ of Chebyshev expansion polynomials is chosen a priori.

2. The unknown expansion coefficients a_k of ψ are written in column form $(a^{(\psi)})$. The corresponding column of the (unknown) function $\psi(\xi_k)$ at at the position of the support points ξ_k, $k = 1, 2, \ldots, N + 1$, is written as $(\psi) = C(a^{(\psi)})$, in which the matrix notation defined in Chap. 4 is used.

3. The functions $F(\xi_k)$ and $V(\xi_k)$ are written as diagonal matrices,

$$F_D = \begin{pmatrix} F(\xi_1) & & & \\ & F(\xi_2) & & \\ & & \ddots & \\ & & & F(\xi_{N+1}) \end{pmatrix}, \quad G_D = \begin{pmatrix} G(\xi_1) & & & \\ & G(\xi_2) & & \\ & & \ddots & \\ & & & G(\xi_{N+1}) \end{pmatrix}.$$

The product $F_D V_D C(a^{(\psi)})$ is a column vector that contains as its entries the quantities $F(\xi_k) V(\xi_k) \psi(\xi_k)$.

4. The expansion coefficients of this column vector are given by the values $C^{-1} F_D V_D C a(a^{(\psi)})$, and the expansion coefficients b_k of $I_R(r) = \int_r^{R_{max}} F V \psi dr'$ are given by $S_R C^{-1} F_D V_D C(a^{(\psi)})$.

5. With the additional multiplication by C the column vector in step 4 is transformed back into a column vector of the corresponding function evaluated at the ξ_k's. That in turn is multiplied by $-(1/k)(R_{max}/2)G_D$, and "voilà", one has the col-

umn vector of the functions representing the term $-\frac{1}{k}G(r)\int_0^r F(r')\,V(r')\,\psi(r')\,dr'$ in Eq. (6.4). That can again be transformed into the column vector of the respective expansion coefficients through an additional multiplication by C^{-1}. The factor $(R_{max}/2)$ is due to the integral being over dr' rather over dx.

In summary, the matrix representation of $-\frac{1}{k}G'(r)\int_0^r F(r')\,V(r')\,\psi(r')\,dr'$ in Eq. (6.4) is

$$-C^{-1}(1/k)(R_{max}/2)G_D C S_L C^{-1} F_D V_D C(a^{(\psi)}).$$

The middle term of Eq. (6.4) is correspondingly

$$-C^{-1}(1/k)(R_{max}/2)F_D C S_R C^{-1} G_D V_D C(a^{(\psi)}),$$

and the sum of these two terms gives rise to the matrix

$$M = -C^{-1}(1/k)(R_{max}/2)\left[F_D C S_R C^{-1} G_D + G_D C S_L C^{-1} F_D\right]V_D C. \quad (6.12)$$

The final equation for the expansion coefficients (a^ψ), according to Eq. (6.4), is

$$(1_D - M)(a^{(\psi)}) = (a^{(F)}). \quad (6.13)$$

In order to solve for $(a^{(\psi)})$, the inverse of the matrix $(1_D - M)$ appears to be required. However in MATLAB the command "\" is used in the form $(a^{(\psi)}) = (1_D - M)\backslash(a^{(F)})$ to solve a set of linear equations, rather than requiring the inverse of that matrix.

The above manipulations are at the heart of the S-IEM method to solve the one-dimensional radial Schrödinger equation. In order to get good accuracy and efficiency, it is desirable to divide the whole radial domain $[0, R_{max}]$ into partitions, using a small number of Chebyshev expansion terms in each partition. A short description is given below and in Chap. 7, and a more extensive description can be found in Refs. [2, 4].

6.3.1 A Numerical Example

For this example the potential is attractive or repulsive, and of a simple exponential nature

$$V(r) = \mp 5 \times \exp(-r), \quad (6.14)$$

the wave number $k = 0.5$, the number of Chebyshev expansion polynomials is $N + 1 = 31$, $L = 0$, and the radial interval is $[0, 12]$. The solution of Eq. (6.4) is obtained by means of the MATLAB code "wave2", listed in the Appendix A. The resulting wave function solutions are shown in Fig. 6.3. For the repulsive case, the coefficient -5 is replaced by 5. The boundary conditions are contained in the Green's function. At the origin the wave functions go to zero, asymptotically they are given by a linear

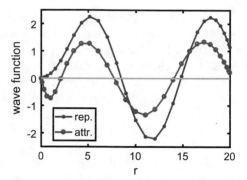

Fig. 6.3 The functions are the solutions of the Lippmann–Schwinger equation (6.4), for $V(r) = (+, -)5 \times \exp(-r)$, in the radial domain $[0, 20]$, with $N = 30$ and $k = 0.5$. This result is obtained with MATLAB code "wave2.m" reproduced in the Appendix. Please note that for the repulsive case the curve lacks the additional minimum near $r = 1$

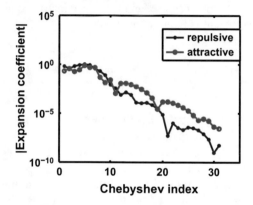

Fig. 6.4 The absolute values of the Chebyshev expansion coefficients for the two functions described in Fig. 6.3. For the repulsive case the expansion coefficients a_n decrease with n faster than for the attractive case. This is because for the repulsive case there is one less minimum to represent

combination of $\sin(kr)$ and $\cos(kr)$ with appropriate coefficients. One sees that for the repulsive case the wave function is "pushed" away (rightward) from the origin, and the minimum near $r \simeq 1$ is suppressed. The computing time is less than 1 s on a desk-top computer Intel TM2 Quad, with a CPU Q 9950, a frequency of 2.83 GHz, and a RAM of 8 GB.

These functions are not normalized to unit amplitude at large distances according to Eq. (6.10), but are given by Eq. (6.8). The absolute value of the Chebyshev expansion coefficients are shown in Fig. 6.4.

According to Fig. 6.4 the accuracy of these attractive and repulsive wave functions is expected to be of the order of 10^{-6} and 10^{-9} respectively, throughout the whole domain. Also to be noted is that the asymptotic amplitude of ψ for the attractive case is approximately 1.4. This amplitude is compatible with Eq. (6.4), which for

$r \simeq R_{max}$ is given by Eq. (6.11). In that expression it is assumed that the value of V beyond R_{max} is so small that it can be neglected to within the desired accuracy. The resulting values obtained with the code "wave" are: $tan(\varphi) = 0.8913331251486368$, and $K = 1.339580061059150$. A more precise calculation performed with a more sophisticated S-IEM code [2, 4], accurate to 10 significant figures, agrees with the value of $tan(\varphi)$ to seven significant figures. Its result is 0.891333117387316. However, if the value of R_{max} was larger, these values would change beyond the 5th significant figure, because in the present example one has $V(R_{max}) \simeq -3 \times 10^{-5}$. The more sophisticated SIEM-method consists in dividing the radial domain into partitions, the length of which is not prescribed but is determined automatically by the size of the last expansion coefficients of two independent functions, and solutions of the L–S equation calculated in the partition. If the size is larger than the specified error parameter, then the partition is divided in half. The solution of Eq. (6.2) is then obtained as a linear combination of the two independent functions described above, and the coefficients of the linear combination are determined by the solution of a matrix equation. That matrix is sparse, and hence the matrix equation can be solved by well established efficient computational methods.

6.3.2   Exercises

1. Starting with Eq. (6.4), prove the validity of Eq. (6.7).
2. Starting with Eq. (6.7), prove the validity of Eq. (6.1).
3. Starting with Eq. (6.4), prove the validity of Eq. (6.11), and theoretically examine the consequences of changing the value of R_{max}.
4. Using the code "wave", examine the stability of the resulting value of T as a function of the number $N + 1$ of Chebyshev polynomials. Keep the values of k, and R_{max} unchanged.
5. Make the potential in code "wave" repulsive (instead of attractive) by replacing the value -5 of V_0 by $+5$. Keep the values of k, and R_{max} unchanged. Give an intuitive interpretation of the result.

6.4   A Shape Resonance

For this case the potential is composed of two exponential functions

$$V_M(r) = V_0 \, e^{-(r-r_e)\,\alpha} \left[2 - e^{-(r-r_e)\,\alpha} \right], \tag{6.15}$$

with

$$V_0 = 4; \, r_e = 4; \, \alpha = 0.3. \tag{6.16}$$

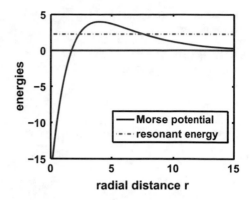

Fig. 6.5 The Morse potential in units of inverse length squared (blue line), given by Eqs. (6.15) and (6.16). A resonant energy is indicated by the dashed green line at $E \simeq 2.25$ inverse length squared

An attractive valley appears near the origin, and a repulsive barrier appears at larger distances, as illustrated in Fig. 6.5. In the legend the potential is denoted as "Morse", because it is the negative of a potential that was first introduced by P.M. Morse in 1930 for the purpose of testing bound state calculations [14].

The corresponding wave functions for various energies in the resonance region are illustrated in Fig. 6.6.

The tangents of their phase shifts are listed in Table 6.1. The wave functions illustrated in Fig. 6.6 do decrease in the barrier region, and the phase shift changes rapidly as a function of energy. This is in contrast to a wave function at a different non-resonant energy, for example when $k = 1.40$. The absolute value of the wave function increases inside of the barrier region, as illustrated in Fig. 6.7, and is much less sensitive to variations in the energy.

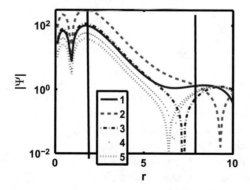

Fig. 6.6 The absolute value of several wave functions corresponding to the energies k^2, with $k = 1.50715 + (n - 1) \times 0.00001$, for $n = 1, 2, \ldots, 5$. The tangents of the corresponding phase shifts are listed in Table 6.1. The two vertical lines delimit the barrier region. Please note that the wave function with largest amplitude to the left of the barrier region occurs for $n = 2$ (green dashed line). All functions are normalized to unit amplitude asymptotically

Fig. 6.7 A wave function
outside of a resonance
region, for $k = 1.4$. In the
barrier region the absolute
value of this wave function
increases with distance

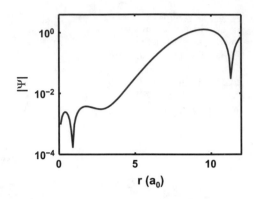

Table 6.1 Information for
Fig. 6.6

n	k	$\tan(\varphi)$	$\varphi(rad)$
1	1.50715	−1.97933	−1.130
2	1.50716	−0.07533	−0.075
3	1.50717	4.81046	1.366
4	1.50718	44.89671	1.548
5	1.50719	−26.00695	1.609

The sensitivity of the wave function and corresponding phase shift with energy
for the resonant case can be understood as follows. In the barrier region there are two
solutions to the Schrödinger equation: one that increases exponentially with distance,
$\psi^{(+)}(r)$, and another, $\psi^{(-)}(r)$, that decreases exponentially, as will be described in
Chap. 8. The complete wave function is given by the linear combination

$$\psi(r) = A\psi^{(+)} + B\,\psi^{(-)}(r), \tag{6.17}$$

where the constants A and B depend on the slope of the wave function in the region
around the left turning point (near 2.2 in the present example). A small change in the
slope due to a change in energy produces a small change in the constants A and B,
making a small difference in ψ near the left turning point. However, near the right
turning point that change is much amplified because at that point $\psi^{(+)}$ is much larger,
and $\psi^{(-)}$ is much smaller than what they are at the left turning point (by about two
orders of magnitude each in the numerical example above, hence the small change is
amplified by four orders of magnitude). Therefore, the wave function to the right of
the right turning point acquires a much changed phase shift. The same effect does not
occur for energies outside of the resonance region, because in this case the decreasing
part of the wave function is much suppressed.

As the energy increases, the wave function is compressed more to the left (towards
the origin), and hence the phase shift increases. This is shown in Fig. 6.8. As the wave
number k increases, as given by $1.50713 + (n − 1) \times 0.5 \times 10^{-5}$ in units of inverse
length, for $n = 1, 2, \ldots, 16$, the phase shift increases approximately by π, starting
from $\simeq -(3/2)\pi$ for $n = 5$ to $\simeq -(1/2)\pi$ for $n = 16$.

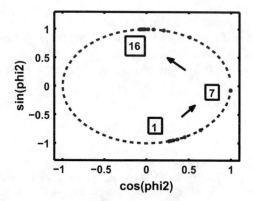

Fig. 6.8 The phase of the wave function as a function of wave number, for the resonance case for the Morse potential described in this chapter. The wave numbers k are given by the expression $k = 1.507130 + (n - 1) \times 0.5 \times 10^{-5}$, in units of inverse length, for $n = 1, 2, \ldots, 16$. Some of the values of n are noted in the figure. The corresponding energies are k^2. The figure shows that as the energy increases, the phase angle increases, moving in the counter-clockwise direction

This resonance has been investigated previously [5]. The accuracy of phase shifts obtained with the S-IEM method is illustrated in Fig. 2 of Ref. [4], and it is compared with the accuracy of a finite difference method. The accuracy of the S-IEM method is higher by six orders of magnitude.

6.4.1 Project

Try to reproduce the results shown in the figures above by using code "wave" as a template. Those results were obtained by using a more elaborate S-IEM code, that automatically divides the radial domain [0, 20] into not equispaced partitions (in this case 12 partitions), whose size is determined by the accuracy parameter 10^{-11}. The radial coordinates of the right and left end of each partition n are denoted as $b1(n)$ and $b2(n)$, $n = 1, 2, \ldots, 12$, respectively. Figure 6.9 displays the right end position

Fig. 6.9 The partition distribution obtained in code S-IEM for the Morse potential resonant case with $k = 1.50715$, in units of inverse length. The rightmost coordinate of each partition n is denoted as $b2(n)$, and is given in units of length

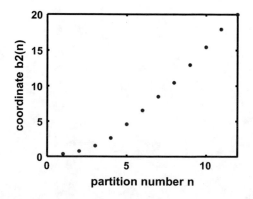

of each partition. The figure indicates that the size of the partitions near the origin are smaller than the size near the end ($r = 20$) of the radial domain. The method is described in Refs. [2, 4]. By contrast, the code "wave" has only one partition in the whole radial interval, and is likely not to give an adequate precision.

References

1. R.H. Landau, *Quantum Mechanics II* (Wiley, New York, 1990)
2. R.A. Gonzales, J. Eisert, I. Koltracht, M. Neumann, G. Rawitscher, Integral equation method for the continuous spectrum radial Schrödinger equation. J. Comput. Phys. **134**, 134–149 (1997)
3. M. Abramowitz, I. Stegun (eds.), *Handbook of Mathematical Functions* (Dover, New York, 1972)
4. G.H. Rawitscher, I. Koltracht, Description of an efficient numerical spectral method for solving the Schroedinger equation. CiSE (Comput. Sci. Eng.) **7**, 58–66 (2005)
5. G. Rawitscher, C. Meadow, M. Nguyen, I. Simbotin, Am. J. Phys. **70**, 935 (2002)
6. A. Gilat, V. Subramaniam, *Numerical Methods for Engineers and Scientists* (Wiley, New York)
7. W.J. Olver, D.W. Lozier, R.F. Boisvert, C.W. Clark, *NIST Handbook of Mathematical Functions*, National Institute of Standards and Technology, U.S. Department of Commerce (Cambridge University Press, Cambridge, 2010)
8. B. Lippmann, J. Schwinger, Variational principles for scattering processes. Phys. Rev. **79**, 469 (1950); B. Lippmann, Phys. Rev. **79**, 481 (1950)
9. N.F. Mott, H.S.W. Massey, *The Theory of Atomic Collisions* (Oxford at the Clarendon Press, London, 1965), starting with Eq. (80) in Chap. IV, Sect. 7
10. R.A. Gonzales, S.Y. Kang, I. Koltracht, G. Rawitscher, Integral equation method for coupled Schrödinger equations. J. Comput. Phys. **153**, 160–202 (1999)
11. G. Rawitscher, A spectral phase-amplitude method for propagating a wave function to large distances. Comput. Phys. Commun. **191**, 33–42 (2015)
12. G. Rawitscher, I. Koltracht, Eur. J. Phys. **27**, 1179 (2006)
13. G.H. Rawitscher, Applications of a numerical spectral expansion method to problems in physics; a retrospective, *Operator Theory: Advances and Applications*, vol. 203 (Birkhauser Verlag, Basel, 2009), pp. 409–426
14. P.M. Morse, H. Feshbach, *Methods of Theoretical Physics*, vol. 2 (McGraw-Hill, New York, 1953), p. 1672

Chapter 7
Spectral Finite Element Method

Abstract In this chapter we describe a method to obtain the solution of second order linear differential equations by means of expansions into sets of Lagrange polynomials called discrete variable representation (DVR). The coefficients of the expansion are obtained by a Galerkin method. When the radial domain is subdivided into contiguous partitions, the total procedure is called the spectral finite element method (FE-DVR). In each partition (or finite element) the desired solution is expanded into a set of basis functions. In the present chapter the basis functions are Lagrange polynomials, and the algorithm is of the Galerkin type. We compare the errors and the speed of calculation of the FE-DVR with those of the S-IEM, where the basis functions are Chebyshev polynomials, and the algorithm is of the Collocation type.

7.1 Summary and Motivation

The main objective of this chapter is to describe a spectral finite element method based on Lagrange polynomials, and compare the efficiency and errors of this method with the S-IEM and finite difference methods described in Chaps. 2 and 6, respectively. All three methods divide the radial domain into partitions but differ in that for the present FE-DVR method the solution in each partition is obtained by a Galerkin procedure, while for the S-IEM method a Collocation method is used. They further differ in that they use distinct algorithms to smoothly propagate the resulting wave function from one partition to the next. Hence they lead to a different accumulation of errors, and hence comparing the performance of these three methods is a very enlightening exercise. The differential equations to be solved are linear and of second order, the numerical case envisaged is that of a Schrödinger one dimensional equation for positive energy scattering situations. The different methods are based on distinct sets of mesh points. The importance of the mesh point distribution is also clarified in this chapter. Since the DVR method has become very popular for scientific and engineering applications, a short historical survey is presented in the introduction.

© Springer Nature Switzerland AG 2018

G. Rawitscher et al., *An Introductory Guide to Computational Methods
for the Solution of Physics Problems*,
https://doi.org/10.1007/978-3-319-42703-4_7

7.2 Introduction

The solution of differential equations by means of expansions into discrete variable representation (DVR) basis functions has become very popular since it was first introduced in the early 1960s [1]. The main idea of the DVR is to expand the wave function in a set of Lagrange polynomials, encompassing the whole radial domain, and obtain the expansion coefficients by a Galerkin method. A review can be found in the paper by Light and Carrington [2], and generalizations to multidimensional expansions are also under development [3–5]. The main purpose of this chapter is to describe a one dimension generalization of this method in the case where the radial domain is subdivided into partitions called finite elements, and in each partition the expansion into Lagrange polynomials is applied. The support points are the Lobatto points, described in Chap. 3. The Galerkin integrals are approximated by discrete sums over the values of the integrand evaluated at the support points times certain weight factors such as those in the Gauss quadrature methods [6], which in this case is called Gauss–Lobatto integration. So, the difference from the more conventional methods is that the basis functions in each partition are not for example "hat" functions, but are spectral Lagrange polynomials. This combination of methods will be denoted as FE-DVR (finite element plus discrete variable representation) in what follows. Hat functions consist of straight lines that make sharp angles with each other, and are easy to calculate, but their derivatives have points of discontinuities, thus introducing numerical errors. In other versions of the FE methods, Gaussian functions are also used. Lagrange functions are defined in Eq. (3.12). They have many computational advantages for calculations using the Galerkin method, since integrals can be carried out very easily and accurately, as explained in Chap. 3 and as shown here. Another very important property of the finite element methods is that they can be generalized to more than one dimension, since the boundary conditions can be incorporated for complicated boundaries [4, 5].

The Lagrange basis set was first suggested by Manolopoulos and Wyatt [7], and an extensive review is given in Ref. [8]. The FE-DVR method has also been used extensively for fluid dynamic calculations since the 1980s [9] and in Seismology [10], where it is called the spectral element method. This method has been introduced into atomic physics by Resigno and McCurdy [11] for quantum scattering calculations, and is now used extensively for atomic physics applications. As mentioned in Chap. 3, the main computational advantage of using Lagrange basis functions is that the Galerkin integrals reduce to only one term, because the product of two different Lagrange functions vanishes at the support points, and only products between the same functions remain. Furthermore, within the approximation of the Gauss–Lobatto quadrature rule, the basis functions are orthogonal. Hence the procedure leads to a discretized Hamiltonian $(N \times N)$ matrix, whose eigenfunctions determine the expansion coefficients and the eigenvalues determine the bound-state energies. This method introduces several types of errors. One type of error arises from the truncation of the expansion of the wave function in terms of basis functions at an upper limit N. Another error is due to the approximation of the Gauss–Lobatto quadrature described previously. A third error is the accumulation of machine round-off errors. These errors have been examined for bound state energy eigenvalues [3, 8, 12]. It can be shown that the convergence of the energy with the number N of basis functions is exponential, and the non-orthogonality error becomes small as N increases. For more information the reader is invited to consult existing textbooks, such as Chapter 9 in Ref. [13]. Chapter 3 of Ref. [14] also presents an extensive discussion, references and examples of the DVR method. However the methods described in that book are different from the one described here.

The main purpose of the present chapter is to analyze the accuracy of the FE-DVR method for the scattering conditions, since all the errors described above (the Gauss–Lobatto's integration error, the truncation errors of the expansions, and the round-off errors) are still present, and to examine how they accumulate. In our study a method of imposing the continuity of the wave function and of the derivative from one partition to the next is explicitly given, and the accuracy is obtained by comparing the results of the FE-DVR calculation for particular solutions of a one dimensional Schrödinger equation with a bench-mark spectral [15] Chebyshev expansion method [16], denoted as S-IEM. The accuracy of the latter is of the order of $1 : 10^{-11}$, as is demonstrated in Ref. [16]. In our present formulation of the FE-DVR the so-called bridge functions used in Ref. [11] for the purpose of achieving continuity of the wave function from one element to the next are not used, but are replaced by another method, described below.

7.3 The Finite Element Method

The FE-DVR version of the finite element method, presented in this chapter, differs from the conventional FEM (Finite Element Method) in that the basis functions for the expansion of the solution $\psi(x)$ in each partition are N "discrete variable representation" (DVR) functions. These in the present case are Lagrange polynomials $\mathscr{L}_i(x)$, $i = 1, 2, \ldots, N$, all of which are of order $N - 1$ as given by Eq. (3.12) in Chapter 3. These functions are widely used for interpolation procedures and are described in standard computational textbooks, for example in Eq. (25.2.2) of Ref. [17], and in section 3.3(i) of Ref. [6]. This FE and DVR combination has the advantage that integrals involving these polynomials amount to sums over the functions evaluated only at the support points. In the present case the support points are Lobatto points x_j and weights w_j, $j = 1, 2, \ldots, N$, as defined in Eq. (25.4.32) of Ref. [17], and also in Chap. 3, in terms of which a quadrature over a function $f(x)$ in the interval $[-1, +1]$ is approximated by

$$\int_{-1}^{+1} f(x)\, dx \simeq \sum_{j=1}^{N} f(x_j) w_j. \tag{7.1}$$

If f is a polynomial of degree $\leq 2N - 3$ then Eq. (7.1), denoted as the Gauss–Lobatto quadrature approximation [18, 19], will be exact. This however is not the case for the product of two Lagrange polynomials $\mathcal{L}_i(x)\mathcal{L}_j(x)$, which is a polynomial of order $2N - 2$. If the integral limits are different from ± 1, such as $\int_a^b f(r)dr$, then the variable r can be scaled to the variable x. A further DVR advantage is that the Gauss–Lobato approximation of the integral

$$\int_{-1}^{1} \mathcal{L}_i(x)f(x)\,\mathcal{L}_j(x)dx \simeq \delta_{i,j}w_j f(x_i) \tag{7.2}$$

is diagonal in i, j and is given by only one term. The convolution

$$\int_{-1}^{1} \mathcal{L}_i(x)\int_{-1}^{1} K\left(x, x'\right)\mathcal{L}_j(x')dx'\,dx \simeq w_i w_j K(x_i, x_j) \tag{7.3}$$

is also approximated by one non-diagonal term only, which is a marked advantage for solving nonlocal or coupled channel Schrödinger equations. Here $K\left(x, x'\right)$ is the kernel for a nonlocal potential case for which the term $V(r)\psi(r)$ is replaced by the integral $\int_0^\infty K(r, r')\psi(r')dr'$. Such a case is examined in Ref. [20], but will not be repeated here.

In applications to the solution of second order differential equations, such as the Scrödinger equation, the integral over the second order derivative operator can be expressed in the form

$$\int_{-1}^{1} \mathcal{L}_i(x)\frac{d^2}{dx^2}\mathcal{L}_j(x)dx = -\int_{-1}^{1} \mathcal{L}_i'(x)\mathcal{L}_j'(x)dx + \delta_{i,N}\mathcal{L}_j'(1) - \delta_{i,1}\mathcal{L}_j'(-1) \tag{7.4}$$

after an integration by parts. In the above the prime denotes d/dx. The integral on the right hand side of this equation can be done exactly with the Gauss–Lobatto quadrature rule (7.1), since the integrand is a polynomial of order $2N - 4$, that is less than the required $2N - 3$. That leads to a very fast computational execution.

For the case of a local potential V with angular momentum number $L = 0$ the equation to be solved is

$$\left(\frac{d^2}{dr^2} + k^2\right)\psi(r) = V(r)\psi(r). \tag{7.5}$$

The wave number k is in units of fm^{-1} and the potential V is in units of fm^{-2}, where quantities in energy units are transformed to inverse length units by multiplication by the well known factor $2m/\hbar^2$. In the scattering case the solutions $\psi(r)$ are normalized such that for $r \to \infty$ they approach

$$\psi(r) \to \sin(kr) + \tan(\delta)\cos(kr), \tag{7.6}$$

and with that normalization one finds

$$\tan(\delta) = T = -\frac{1}{k} \int_0^\infty \sin(kr)\, V(r)\, \psi(r) dr, \tag{7.7}$$

as is well known [21].

As described in Ref. [22], the FE-DVR procedure is as follows. We divide the radial interval into N_J partitions (also called elements in the finite element calculations [23]), and in each partition (J), with $J = 1, 2, \ldots, N_J$, we expand the wave function into N Lagrange functions $\mathscr{L}_i(r)$, $i = 1, 2, \ldots, N$,

$$\psi^{(J)}(r) = \sum_{i=1}^N c_i^{(J)} \mathscr{L}_i(r), \quad b_1^{(J)} \le r \le b_2^{(J)}, \tag{7.8}$$

where the starting and end points of each partition are denoted as $b_1^{(J)}$ and $b_2^{(J)}$, respectively. The superscript (J) indicates that the quantities refer to partition J.

By performing the Galerkin integrals of the Schrödinger equation over the $\mathscr{L}_i$ in each partition J

$$\left(\mathscr{L}_i | (T + V - k^2) \psi^{(J)}\right) =$$
$$= \int_{b_1^{(J)}}^{b_2^{(J)}} \mathscr{L}_i(r)(T + V - k^2)\psi^{(J)}(r) dr = 0, \quad i = 1, 2, \ldots, N, \tag{7.9}$$

and after expanding $\psi^{(J)}$ into Lagrange polynomials given by Eq. (7.8), we obtain a homogeneous matrix equation in each partition for the coefficients $c_i^{(J)}$, $i = 1, 2, \ldots, N$,

$$\sum_{j-1}^N M_{ij}^{(J)} c_j^{(J)} = 0, \tag{7.10}$$

where the matrix elements of M are given by $M_{ij}^{(J)} = \left(\mathscr{L}_i | (T + V - k^2)\mathscr{L}_j\right)$, and where $T = -d^2/dr^2$.

7.3.1 The Continuity Conditions

The two continuity conditions are imposed by transforming the homogeneous Eq. (7.10) of dimension N into an inhomogeneous system of equations of dimension $N - 2$ whose driving terms are composed of the function ψ and $d\psi/dr$ evaluated at the end of the previous partition. We obtain the value of the wave function at the end point of the previous partition as

$$\psi^{(J-1)}|_{b_2^{(J-1)}} = c_N^{(J-1)}, \tag{7.11}$$

where $c_N^{(J-1)}$ is the last coefficient of the expansion (7.8) of $\psi^{(J-1)}$, and where the subscript $|_{b_2^{(J-1)}}$ means the the function $\psi^{(J-1)}$ is being evaluated at the point $b_2^{(J-1)}$. The derivative of the wave function at the right end of the previous partition is given by

$$A^{(J-1)} \equiv \frac{d}{dr} \psi^{(J-1)}|_{b_2^{(J-1)}} = \sum_{i=1}^{N} c_i^{(J-1)} \mathcal{L}'_i|_{b_1^{(J-1)}}, \qquad (7.12)$$

respectively, where $\mathcal{L}'_i(r) = d\mathcal{L}_i(r)/dr$. The result (7.11) follows from the fact that that $\mathcal{L}_i(b_2) = 0$ for $i = 1, 2, \ldots, N-1$, and $\mathcal{L}_N(b_2) = 1$. For the first partition we arbitrarily take a guessed value of $A^{(0)}$ for the non-existing previous partition, and later renormalized the whole wave function by comparing it to a known value. That is equivalent to renormalizing the value of $A^{(0)}$.

Hence, the two continuity conditions of the wave function from one partition $(J-1)$ to the next (J) are

$$c_1^{(J)} = c_N^{(J-1)}, \qquad (7.13)$$

where we use $\mathcal{L}_i(b_1) = 0$ for $i = 2, \ldots, N$, and $\mathcal{L}_1(b_1) = 1$, and

$$\frac{d\,\psi^{(J-1)}}{dr}|_{b_2^{(J-1)}} = \sum_{i=1}^{N} c_i^{(J)} \mathcal{L}'_i|_{b_1^{(J)}} = A^{(J-1)}. \qquad (7.14)$$

These two conditions can be written in the matrix form

$$F_{11}\alpha + F_{12}\beta = \gamma, \qquad (7.15)$$

where

$$F_{11} = \begin{pmatrix} 1 & 0 \\ \mathcal{L}'_1 & \mathcal{L}'_2 \end{pmatrix}^{(J)}_{b_1^{(J)}}; \quad F_{12} = \begin{pmatrix} 0 & 0 & \cdots & 0 \\ \mathcal{L}'_3 & \mathcal{L}'_4 & \cdots & \mathcal{L}'_N \end{pmatrix}^{(J)}_{b_1^{(J)}}, \qquad (7.16)$$

where

$$\alpha = \begin{pmatrix} c_1 \\ c_2 \end{pmatrix}^{(J)}, \qquad (7.17)$$

where

$$\beta = \begin{pmatrix} c_3 \\ c_4 \\ \vdots \\ c_N \end{pmatrix}^{(J)}, \qquad (7.18)$$

and where

$$\gamma = \begin{pmatrix} c_N \\ A \end{pmatrix}^{(J-1)}. \qquad (7.19)$$

With that notation, Eq. (7.10) can be written in the form

$$\begin{pmatrix} M_{11} & M_{12} \\ M_{21} & M_{22} \end{pmatrix} \begin{pmatrix} \alpha \\ \beta \end{pmatrix} = 0, \tag{7.20}$$

where the matrix $M^{(J)}$ has been decomposed into four sub-matrices M_{11}, M_{12}, M_{21}, and M_{22}, which are of dimension 2×2, $2 \times (N - 2)$, $(N - 2) \times 2$, and $(N - 2) \times (N - 2)$, respectively. The column vector α can be eliminated in terms of β and γ by using Eq. (7.15)

$$\alpha = F_{11}^{-1}(-F_{12}\beta + \gamma), \tag{7.21}$$

and the result when introduced into Eq. (7.20) leads to an inhomogeneous equation for β

$$(-M_{21} F_{11}^{-1} F_{12} + M_{22})\beta = -M_{21} F_{11}^{-1} \gamma. \tag{7.22}$$

Once the vector β is found from Eq. (7.22), then the components of the vector α can be found from Eq. (7.21), and the calculation can proceed to the next partition.

If one expresses the inverse of F_{11} analytically

$$F_{11}^{-1} = \begin{pmatrix} 1 & 0 \\ -\dfrac{\mathscr{L}_1'}{\mathscr{L}_2'} & \dfrac{1}{\mathscr{L}_2'} \end{pmatrix}. \tag{7.23}$$

then one finds

$$F_{11}^{-1}\gamma = \begin{pmatrix} c_N^{(J-1)} \\ -\dfrac{\mathscr{L}_1'}{\mathscr{L}_2'} c_N^{(J-1)} + \dfrac{A^{(J-1)}}{\mathscr{L}_2'} \end{pmatrix} \tag{7.24}$$

and

$$F_{11}^{-1} F_{12} = \begin{pmatrix} 0 & 0 & \cdots & 0 \\ \dfrac{\mathscr{L}_3'}{\mathscr{L}_2'} & \dfrac{\mathscr{L}_4'}{\mathscr{L}_2'} & \cdots & \dfrac{\mathscr{L}_N'}{\mathscr{L}_2'} \end{pmatrix}. \tag{7.25}$$

By inserting (7.23) into (7.21) one finds that $c_1^{(J)} = c_N^{(J-1)}$, but $c_2^{(J)}$ is a function of $c_N^{(J-1)}$, $A^{(J-1)}$, and the vector β.

7.4 Accuracy

In what follows, we test the accuracy for a local potential V_M with angular momentum $L = 0$. Potential V_M is of a Morse type given by

$$V_M(r) = V_0 \, e^{-(r-r_e)\,\alpha} \left[2 - e^{-(r-r_e)\,\alpha}\right],$$
$$V_0 = -6; \ r_e = 4; \ \alpha = 0.3, \tag{7.26}$$

Fig. 7.1 The Morse potential V_M as a function of radial distance r, in units of inverse length squared. This potential is given by Eq. (7.26)

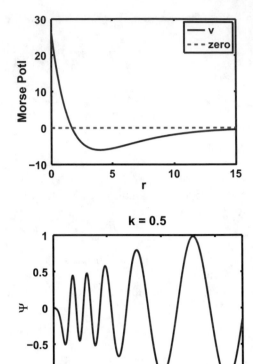

Fig. 7.2 The wave function for the Morse potential, at a wave number $k = 0.5$. It is normalized to unit amplitude at $r > 100$

and illustrated in Fig. 7.1. It has a repulsive core near the origin and decays exponentially at large distances. The coefficient -6 is in units of fm^{-2}, the distances r are in units of fm, and all other factors are such that the arguments of the exponents are dimensionless. This potential is the negative of the barrier type potential used for a resonance calculation in Chap. 6. The wave function for the Morse potential at a wave number k = 0.5, normalized to unit amplitude at r > 100, is illustrated in Fig. 7.2.

In order to obtain an accuracy of $1 : 10^{-11}$ the bench-mark S-IEM wave function has to be calculated out to 100 fm. Beyond this point the magnitude of the potential is less than 10^{-11}.

In order to ascertain the accuracy of the FE-DVR method, the solutions of Eq. (7.5) are compared with the solutions obtained by the spectral integral equation method (S-IEM) [16], with an accuracy of $1 : 10^{-11}$, as described in Chap. 6. The numerical FE-DVR solutions are first normalized by comparison with the S-IEM solutions at one chosen radial position near the origin, and the error of the normalized FE-DVR function is determined by comparison with the S-IEM function at all other radial points r. Since the S-IEM function depends on the values of the potential at all points $[0 \leq r \leq R_{max}]$, the S-IEM calculation has to be carried out to a distance

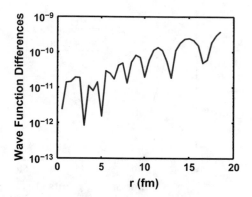

Fig. 7.3 The accuracy of the FE-DVR wave function for the potential V_M as a function of the size of the radial domain. The accuracy is obtained by comparison with the S-IEM result, which is accurate to $1:10^{-11}$ for all sizes of the radial domain. The wave number is $k = 0.5\,\text{fm}^{-1}$, the number of Lobatto points per partition is 20, and the size of each partition is 1 fm

$R_{max} = 100\,\text{fm}$ where $|V| \leq 10^{-11}$. This distance is large enough so that the contribution from $V(r \geq R_{max})$ is smaller than the desired accuracy of the S-IEM solution. The same is not the case for the FE-DVR solutions $\psi_{FE-DVR}(r)$, since the un-normalized solution depends only on the potentials for distances less than r. However, if the normalization of the wave function (7.6) is to be accomplished by matching it to $\sin(kr)$ and $\cos(kr)$ at R_{max} in the asymptotic region, then the numerical errors that accumulate out to R_{max} will affect the wave function at all distances in terms of the error of the normalization factor.

The accuracy of the wave function for potential V_M as a function of the length r of the total radial domain is shown in Fig. 7.3. The number of Lobatto points in each partition is the same, and the length of each partition is also the same, as described in the caption of the figure. The error starts with 10^{-11} at the small lengths, and increases exponentially to 10^{-10} as the length of the domain increases due to the accumulation of various errors, as is explained in Appendix B of Ref. [22].

The accuracy as a function of the number N of Lobatto points in each partition, for a fixed size of all partitions and a fixed length of the radial domain, is shown in Fig. 7.4. The accuracy is expressed in terms of the accuracy of the phase shift, given by the integral (7.7).

The open circles represent an upper limit of the estimated accuracy as developed in Appendix B, of Ref. [22]. This figure shows the nearly exponential increase of accuracy as N increases (for the first three points), until the accumulation of errors overwhelms this effect once the value of N increases beyond a certain value. Thus, for this particular example the optimum number of Lobatto points in each partition is 20. If the length of each partition is made larger without at the same time increasing the number of Lobatto points in each partition, then the accuracy deteriorates exponentially, as shown in Fig. 7 of Ref. [22].

A comparison between the FE-DVR and a finite difference sixth order Numerov method in terms of the accuracy of $\tan(\delta)$ is illustrated in Fig. 7.5. The FE-DVR results show that the accuracy of the phase shift (and hence that of the wave function)

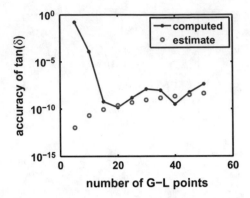

Fig. 7.4 Accuracy of the integral $\int_0^{100} \sin(kr)\, V_M(r)\, \psi(r)\, dr$, obtained with the FE-DVR method as a function of the number of Lobatto points in each partition. The length of each partition is 1.0 fm, the number of partitions is 100. The potential is V_M, the wave number is $k = 0.5\,\text{fm}^{-1}$. The accuracy is obtained by comparison with the S-IEM result which is accurate to $1:10^{-11}$. The open circles represent an estimate of the upper bound for the accumulation of roundoff errors, given in Appendix B of Ref. [22]

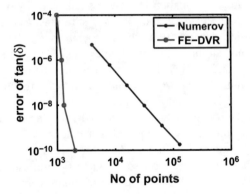

Fig. 7.5 This accuracy comparison for $\tan(\delta)$ is performed for the potential V_M and $k = 0.5\,\text{fm}^{-1}$ in the radial domain [0, 100 fm]. The partition sizes in the FE-DVR method have a length of 1 fm each, and the number of Lobatto points in each partition is given by 1/100th of the total number of points. Numerov is a 6th order finite difference method with equidistant points, mentioned in Chap. 2. The Numerov method requires 10^5 points in order to achieve an accuracy comparable to the FE-DVR method with 2×10^3 points

increases very rapidly as the number of the Lobatto points in each partition increases, (in agreement with Fig. 7.4) much more rapidly than the increase of the accuracy with the number of Numerov points.

This comparison also shows that for an accuracy of $\tan(\delta)$ of $\simeq 10^{-8}$, the FE-DVR method requires 15 times fewer mesh points, and is approximately 100 times faster than the Numerov method.

Finally, the FE-DVR computing time as a function of the number N of Lobatto points in each partition is displayed in Fig. 7.6, where it is also compared with an

Fig. 7.6 The computing time in MATLAB for the calculations described in Fig. 7.4. The dashed line represents the computing time for the S-IEM calculation, described in Figs. 7.6 and 7.7. The open circles represent a estimate described in Ref. [22]

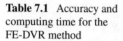

Table 7.1 Accuracy and computing time for the FE-DVR method

# of Pts.	Err[tan(δ)]	Time (s)
2000	10^{-10}	0.075
1300	10^{-8}	0.050
1200	10^{-6}	0.047
1000	10^{-4}	0.045
700	10^{-2}	0.042

estimate described in Appendix B of Ref. [22] based on the number of floating point operations expected. The MATLAB computations are performed on a desktop using an Intel TM2 Quad, with a CPU Q 9950, a frequency of 2.83 GHz, and a RAM of 8 GB. The dashed line represents the total time required for a comparable S-IEM computation. That comparison shows that the FE-DVR method can be substantially faster than the S-IEM, depending on the radial range and on the accuracy required, even though the former has many more support points (Table 7.1).

7.5 Numerical Comparison with the S-IEM Method

As described in Chap. 6, the solution of the Schrödinger equation can be obtained by solving Lippmann–Schwinger (L–S) integral equation. In order to obtain high accuracy and computational efficiency this L–S method can be implemented by dividing the total radial interval into partitions (or finite elements), and expanding the solution in each partition in terms of Chebyshev polynomials. This method has been described in Ref. [16], and a pedagogical version is found in Ref. [24], so the minute details will not be described here. The method consists in obtaining two independent solutions of the L–S equation (and correspondingly also of the Schrödinger Eq. (7.5)) in each partition J, denoted as $Y^{(J)}(x)$ and $Z^{(J)}(x)$, mapped to the interval $[-1, +1]$. The corresponding discretized matrices are not sparse, but are of a small dimension equal to the number of Chebyshev points per partition (of the order of 17 or 30).

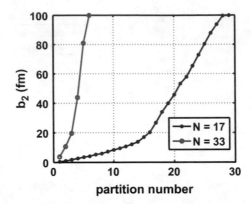

Fig. 7.7 An illustration of the adaptive procedure of the S-IEM method showing the partition distribution in the radial interval [0, 100 fm] for two different numbers $N + 1$ of the Chebyshev expansion functions in each partition. The end point $b_2^{(J)}$ of each partition J is shown on the vertical axis, and the corresponding partition number is shown on the horizontal axis. The calculation using 18 Chebyshev points in each partition requires a larger number of partitions than the calculation with 33 points in each partition in order to achieve the same accuracy, although both calculations require approximately the same computation time. The potential is V_M, Eq. (7.26), and the wave number is $k = 0.5\,\mathrm{fm}^{-1}$. The accuracy parameter *tol* in each partition is 10^{-12}. The computation time for each case is approximately the same, 0.2 s, and the accuracy of the wave function in both cases is also approximately the same, $1 : 10^{-11}$

The solution $\psi^{(J)}$ in each partition is obtained by a linear combination of the two independent functions $Y^{(J)}(x)$ and $Z^{(J)}(x)$, with coefficients that are determined from the solution of a matrix equation of dimension twice as large as the number N_J of partitions. However, the corresponding matrix is sparse.

One of the interesting features of this finite element spectral method (S-IEM) is that the size of each partition is adaptively determined such that the accuracy of the functions $Y^{(J)}(x)$ and $Z^{(J)}(x)$ is equal or better than a pre-determined accuracy parameter *tol*, which in the present case is *tol* $= 10^{-12}$. This adaptive method is based on the size of the last expansion coefficient, which is required to be equal or less than the value of *tol*. If this requirement is not satisfied, the size of the interval is divided in half, and the process continues using in each new partition the same number $N + 1$ of Chebyshev polynomials. In the region where the potential V is small the corresponding partition size is large. In the radial region where the potential is large, the wave function has a smaller local wave length, and the size of the partition becomes smaller automatically. This is illustrated in Fig. 7.7, where N increases from 17 to 33 the number of partitions decreases from 29 to 6. Yet the accuracy of the respective wave functions is approximately the same, $1 : 10^{-11}$, and the computing time is also approximately the same, 0.2 s, as is demonstrated in Table 7.2.

For the present S-IEM benchmark calculations the value of N is 17, and for the case of V_M the maximum value of r is 100 fm. Such a large value is required because the potential decays slowly with distance and becomes less in magnitude than 5×10^{-12}

Table 7.2 Accuracy and computing time for the S-IEM method

Tol.	Part'ns	Points	Err[tan(δ)]	Time (s)
10^{-12}	37	629	–	0.178
10^{-10}	25	425	4.6×10^{-12}	0.181
10^{-8}	17	289	7.7×10^{-11}	0.171
10^{-6}	11	187	5.2×10^{-7}	0.165
10^{-4}	7	119	2.8×10^{-4}	0.162
10^{-2}	5	85	6.5×10^{-2}	0.161

only beyond $r = 100$ fm. Had the potential been truncated at a smaller value of r, then the truncation error would have propagated into all values of the wave function and rendered it less accurate everywhere. The accuracy of the S-IEM wave function can be gauged by comparing two S-IEM wave functions with different accuracy parameters $tol = 10^{-11}$ and 10^{-12}, respectively. The result is that the accuracy of the IEM wave function for $N = 17$ and $R_{\max} = 100$ fm and $tol = 10^{-11}$ is 4×10^{-11}, and that for $tol = 10^{-12}$ the accuracy is better than 10^{-11}. In the case of the resonance example given in Chap. 6, the phase shifts could be calculated analytically and compared with the numerical results. For energies outside the resonance region the accuracy was $\simeq 10^{-12}$, while in the resonance region it was $\simeq 10^{-10}$, as demonstrated in Fig. 2 of Ref. [25].

If the tolerance parameter is reduced gradually from 10^{-12} to 10^{-2} the number of partitions decreases correspondingly, and so does the accuracy of the phase shift, as is shown in Table 7.2.

The error of $\tan(\delta)$ is obtained by comparing the value of $\tan(\delta)$ for a particular tolerance parameter with the value obtained for $tol = 10^{-12}$. The number of Chebyshev polynomials in each partition is 17, and the total number of points displayed in the third column is equal to 17 times the number of partitions.

For the case of a nonlocal potential, $V(r)\psi(r)$ is replaced by $\int_0^r K(r, r')\psi(r')dr'$ in the Schrödinger equation. In this case the division of the radial interval into partitions is not made because the effect of the nonlocal potential would extend into more than one partition, making the programming more cumbersome. For the case of a kernel $K(r, r')$, described in Ref. [20], the accuracy of the S-IEM result [20] is also good to $1 : 10^{-11}$, as is shown in Fig. 7 of Ref. [20].

7.6 Numerical Comparison with a Finite Difference Method

The finite difference method used for this comparison is Milne's corrector method, also denoted as the Numerov method, given by Eq. (25.5.21C) in Ref. [17]. In this method the error of the propagation of the wave function from two previous points to the next point is of order h^6, where h is the radial distance between the consecutive

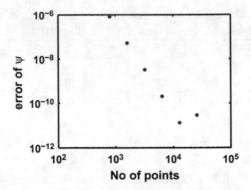

Fig. 7.8 The error of the Numerov wave function at $r = 18$ fm, as a function of the number N of mesh points in the interval $[0, 20$ fm$]$. The distance h between points is $20/N$. For each h the wave function is normalized to the S-IEM wave function at $r = 2$ fm. The wave number is $k = 0.5$ fm^{-1}, the potential is V_M. For a number of points larger that 10^4 (last point on the graph) the accumulation of roundoff errors overwhelms the algorithm error

Table 7.3 Accuracy and computing time for the Numerov method.

N of Pts.	Err[tan(δ)]	Time (s)
12800	1.23×10^{-9}	51
6400	9.41×10^{-9}	5.8
3200	7.50×10^{-8}	2.1
1600	5.99×10^{-7}	1.0
800	4.76×10^{-6}	0.72

equispaced points. The calculation is done for the potential V_M and for $k = 0.5$ fm^{-1} as follows.

A value of h is selected and the Milne wave function is calculated starting at the two initial points $r = h$ and $r = 2h$ by a power series expansion of the wave function for the potential V_M. The values of the wave function for the additional points $3h, 4h, \ldots$ are obtained from Milne's method out to the point $r = 20$ fm. The wave function is normalized to the S-IEM value at $r = 2$ fm, and the error at $r = 18$ fm is obtained by comparison with the S-IEM value at that point. The result for a sequence of h values is illustrated in Fig. 7.8. The slope of the points in his figure shows the rate of the increase of the accuracy with a decrease in h, and it also shows that for approximately 10^4 Numerov points the accumulation of numerical round-off errors overwhelms the increase of accuracy due to a reduction of h.

For each value of h the wave function is calculated out to $r = 100$ fm by Numerov's method, and the integral (7.7) is calculated by the extended Simpson's rule, given by Eq. (25.4.6) in Ref. [17]. The error is determined by comparison with the S-IEM result 2.6994702502 for tan(δ). More detail of the error and the computing time for the Numerov method is displayed in Table 7.3.

The calculation is done in MATLAB performed on a desktop using an Intel TM2 Quad, with a CPU Q 9950, a frequency of 2.83 GHz, and a RAM of 8 GB.

7.7 Summary and Conclusions

The accuracy of a hybrid finite element method (FE-DVR) has been examined for the solution of the one dimensional Schrödinger equation with scattering boundary conditions. This method [11] uses as basis functions the discrete variable representation Lagrange polynomials $\mathscr{L}_i(r)$, $i = 1, 2, \ldots, N$, on a mesh of N Lobatto support points. The accuracy of the FE-DVR method is obtained by comparison with a spectral finite element method S-IEM, whose accuracy is of the order of $1 : 10^{-11}$. An important advantage of a discrete variable representation basis is the ease and accuracy with which integrals can be performed using a Gauss–Lobatto integration algorithm that furthermore renders the matrix elements $\left(\mathscr{L}_i|(V - E)\mathscr{L}_j\right)$ diagonal. This feature also permits one to easily solve the Schrödinger equation in the presence of nonlocal potentials with a kernel of the form $K(r, r')$, as is demonstrated in Ref. [26]. Another advantage is that the Galerkin matrix elements of the kinetic energy operator T need not be recalculated anew for each partition because they are the same in all partitions to within a normalization factor that only depends on the size of the partition. A further advantage is that the convergence of the expansion (7.8) with the number N of basis functions is exponential, in agreement of what is the case for bound state finite element calculations with Lobatto discretizations [27]. A possible disadvantage may be that if the number of the Lagrange polynomials in each partition is very large and/or the number of partitions is large, as is the case for long ranged potentials, then the accumulation of round-off and algorithm errors may become unacceptably large. In this case a different technique for solving the Schrödinger equation may be required [26, 28].

A review of the many figures presented in this chapter is as follows. They all refer to the same numerical case based on a Morse-like potential:

1. Figure 7.2 shows the wave function for a particular numerical example in order to illustrate the larger number of oscillations near the origin;
2. Figure illustrates the accumulation of the wave function error of the FE-DVR method as a function of distance from the origin;
3. Figure 7.4 shows how the wave-function error decreases as the number of basis functions in each partition increases, but after this number becomes greater than 20, the accumulation of round-off errors is larger than the reduction of the algorithm error;
4. Figure 7.5 compares the accuracy of the FE-DVD calculation with a finite difference Numerov calculation. It shows that the FE-DVD method is far more economical that the Numerov method;
5. Figure 7.6 compares the computation time of the FE-DVD method with the time for a S-IEM method. For a limited size of the radial domain, the FE-DVD method is several times faster;
6. Figure 7.7 illustrates a useful automatic adaptive capability of the S-IEM method, by showing that as the number N of Chebyshev polynomials in each partition increases, the size of each partition can become larger, for the same required

accuracy. This is demonstrated by showing the upper radial value of each partition as a function of the partition number for two values of N;

7. Finally, Fig. 7.8 shows the error of the wave-function as calculated with the Numerov method. As the distance h between successive points decreases, and the corresponding number of support points increases, the error decreases until the accumulation of round-off errors overwhelms the gain in accuracy of the algorithm error, which in this case is of the order $\mathcal{O}(h^4)$.

In summary, for scattering solutions of the Schrödinger equation the accuracy of the FE-DVR method increases exponentially with the number of Lagrange polynomials in each partition until the accumulation of round-off errors overwhelm the result algorithm error. The FE-DVR can easily achieve an accuracy of the order of 10^{-10} for the scattering phase shifts for either local or nonlocal short ranged potentials. In this case the calculation is less complex than the spectral finite element S-IEM method, and is faster than the S-IEM method if the radial domain is sufficiently limited in size. In addition, it is substantially more efficient than a finite difference Numerov method. The latter result is demonstrated by the fact that the FE-DVR was found to be a hundred times faster than the Numerov for an accuracy of 10^{-8} in the scattering phase shift.

References

1. D.O. Harris, G.G. Engerholm, W.D. Gwinn, J. Chem. Phys. **43**, 1515 (1965); A.S. Dickinson, P.R. Certain, J. Chem. Phys. **49**, 4209 (1968)
2. J.C. Light, T. Carrington Jr., Adv. Chem. Phys. **114**, 263 (2000)
3. R.G. Littlejohn, M. Cargo, T. Carrington Jr., K.A. Mitchell, B. Poirer, J. Chem. Phys. **116**, 8691 (2002)
4. G. Sewell, Adv. Eng. Softw. **41**(5), 748–753 (2010)
5. T. Atkins, M. Escudier, *A Dictionary of Mechanical Engineering* (Oxford University Press, Oxford, 2013). ISBN: 9780199587438; E-ISBN: 9780191752308
6. F.W. Olver, D.W. Lozier, R.F. Boisvert, C.W. Clark, *NIST Handbook of Mathematical Functions* (National Institute of Standards and Technology and Cambridge University Press, New York, 2010)
7. D.E. Manolopoulos, R.E. Wyatt, Chem. Phys. Lett. **152**, 23 (1988)
8. B.I. Schneider et al., in *Quantum Dynamic Imaging: Theoretical and Numerical Methods.* CRM Methods in Mathematical Physics, ed. by A.D. Bandrauk, M. Ivanov (Springer Science and Business Media, New York, 2011), p. 149
9. A.T. Patera, J. Comput. Phys. **54**, 468 (1984)
10. J. Tromp, D. Komatitsch, Q. Liu, Commun. Comput. Phys. **3**, 1 (2008)
11. T.N. Rescigno, C.W. McCurdy, Phys. Rev. A **62**, 032706 (2000)
12. H. Wei, J. Chem. Phys. **106**, 6885 (1997)
13. C.F. Gerald, P.O. Wheatley, *Applied Numerical Analysis*, 7th edn. (Pearson, Addison Wesley, Boston, 2004)
14. B.D. Shizgal, *Spectral Methods in Chemistry and Physics. Applications to Kinetic Theory and Quantum Mechanics* (Springer, Dordrecht, 2015)
15. L.N. Trefethen, *Spectral Methods in MATLAB* (SIAM, Philadelphia, 2000); J.P. Boyd, *Chebyshev and Fourier Spectral Methods*, 2nd revised edn. (Dover Publications, Mineola, 2001); B. Fornberg, *A Practical Guide to Pseudospectral Methods.* Cambridge Monographs on Applied and Computational Mathematics (Cambridge University Press, Cambridge, 1998)

16. R.A. Gonzales, J. Eisert, I. Koltracht, M. Neumann, G. Rawitscher, J. Comput. Phys. **134**, 134–149 (1997); R.A. Gonzales, S.-Y. Kang, I. Koltracht, G. Rawitscher, J. Comput. Phys. **153**, 160–202 (1999); G. Rawitscher, I. Koltracht, Comput. Sci. Eng. **7**, 58 (2005); G. Rawitscher, Applications of a numerical spectral expansion method to problems in physics: a retrospective, in *Operator Theory, Advances and Applications*, vol. 203, ed. by T. Hempfling (Birkäuser Verlag, Basel, 2009), pp. 409–426; A. Deloff, Ann. Phys. (NY) **322**, 1373–1419 (2007)

17. M. Abramowitz, I. Stegun (eds.), *Handbook of Mathematical Functions* (Dover, New York, 1972)

18. Z. Kopal, *Numerical Analysis* (Wiley, New York, 1961)

19. V.I. Krylov, *Approximate Calculation of Integrals* (MacMillan, New York, 1962)

20. G.H. Rawitscher, Nucl. Phys. A **886**, 1 (2012)

21. R.H. Landau, *Quantum Mechanics II* (Wiley, New York, 1990)

22. J. Power, G. Rawitscher, Accuracy of a hybrid finite-element method for solving a scattering Schrodinger equation. Phys. Rev. E **86**, 066707 (2012)

23. K.J. Bathe, E. Wilson, *Numerical Methods in Finite Element Analysis* (Prentice Hall, Englewood Cliffs, 1976); O.C. Zienkiewicz, *The Finite Element Method: Its Basis and Fundamentals* (Elsevier Butterworth-Heinemann, Oxford, 2005)

24. G. Rawitscher, J. Liss, Am. J. Phys. **79**, 417–427 (2011)

25. G. Rawitscher, I. Koltracht, Comput. Sci. Eng. **7**, 58 (2005)

26. G. Rawitscher, Iterative evaluation of the effect of long-range potentials on the solution of the Schrödinger equation. Phys. Rev. A **87**, 032708 (2013)

27. M.J. Rayson, Phys. Rev. E **76**, 026704 (2007)

28. G. Rawitscher, Comput. Phys. Commun. **191**, 33–42 (2015)

Chapter 8
The Phase-Amplitude Representation of a Wave Function

Abstract In this chapter, we describe the Phase-Amplitude Method (Ph-A) for the representation of the solution of a Schrödinger equation, in which the wave function is described in an efficient way by its amplitude $y(r)$ and the wave phase $\phi(r)$. Since each of these quantities vary monotonically and slowly with distance, they are much easier to calculate than the wave function itself. An iterative method to solve the non-linear equation for y is described, and the region of convergence of the iterations is examined for two scattering cases: (a) when the potential is smaller than the incident energy (in this case the wave function is oscillatory); and (b) when the potential is larger than the incident energy (this is the case of the classically forbidden region). Various applications and their accuracies are also described in this chapter.

8.1 Summary and Motivation

The purpose of this chapter is to present a very efficient way to represent a one-dimensional wave function in coordinate space in terms of two r dependent quantities: the amplitude $y(r)$ and the phase $\phi(r)$ of the wave. These quantities vary much more slowly with distance than the wave itself, hence the amount of memory they require is much less, and the calculation requires far fewer mesh points. Although the application is for the solution of a Schrödinger equation, applications to compress the information for transmission of sound or visual images could also emerge from this phase-amplitude (Ph-A) method. The calculations reported in this chapter are done by means of a spectral Chebyshev expansion, which is ideally suited for this case, since for this slowly varying function y the whole radial domain needs not to be divided into partitions (comparisons with other expansion functions have not done here; however, another different spectral method is presented in Chap. 12). A draw-back is the fact that the differential equation for $y(r)$ is non-linear. In addition, many physicists still pay close attention to an iterative solution first described by Seaton and Peach in 1962 [1]. We examine two cases here. In one case where the wave function is oscillatory, the potential is smaller that the energy. In the other case

© Springer Nature Switzerland AG 2018

G. Rawitscher et al., *An Introductory Guide to Computational Methods for the Solution of Physics Problems*, https://doi.org/10.1007/978-3-319-42703-4_8

where the wave function varies exponentially with distance, the potential is larger than the energy. The latter occurs in the classically forbidden region. Each case is treated in a separate section.

8.2 Introduction

For attractive potentials the Phase-Amplitude (Ph-A) representation of a wave function $\psi(r)$ is written as

$$\psi(r) = y(r)\sin[\phi(r)], \tag{8.1}$$

where y and φ are the amplitude and phase respectively, while for repulsive potentials the term $\sin[\phi(r)]$ is replaced by a combination of exponentials, increasing or decreasing with distance. The phase and amplitude (Ph-A) of a wave function $\psi(r)$ vary slowly with distance, in contrast to the wave function that can be highly oscillatory. Hence the calculation of the phase and or the amplitude requires far fewer computational mesh points than the wave function itself, and hence is very desirable. In 1930 Milne [2] presented an equation for the phase and the amplitude functions (which is different from the one developed by Calogero [3]), but the difficulty is that the equation for $y(r)$ is non-linear, and hence more difficult to solve than for linear differential equations. In 1960 Seaton and Peach [1] demonstrated a method to solve this non-linear equation iteratively. The objective of the present chapter is to show how to improve Seaton and Peach's iteration procedure with a spectral Chebyshev expansion method, and at the same time to present a non-iterative analytic solution to an approximate version of the iterative equations which is computationally very fast.

This chapter is divided into two parts: one, dedicated to attractive potentials where the wave function oscillates sinusoidally at large distances, while the other part is dedicated to repulsive potentials where the wave function changes by a combination of exponentials. Two numerical examples are given for the attractive case. The first is for a potential $V(r)$ that decreases with distance as $1/r^3$. The second is a Coulomb potential $\propto 1/r$. In the latter case the whole radial range of $[0, 2000]$ requires only between 25 and 100 mesh points and the corresponding accuracy ranges between 10^{-3} and 10^{-6}. The 0th iteration is identical to the Wentzel Kramers Brillouin (WKB) approximation [4, 5].

The region of validity of the WKB approximation requires that the local wave length $\lambda(r) = 2\pi k/[E - V(r)]$ changes little in the distance of the local wave length, i. e., that

$$\frac{\Delta\lambda}{\lambda} = \frac{2\pi k}{(V - E)^2}\frac{dV}{dr} << 1. \tag{8.2}$$

It is found that the conditions of applicability of the Ph-A method are similar. As it will be shown, the WKB approximation gives an accuracy of 10^{-2}, and the 8th iteration improves the accuracy to 10^{-6}. This spectral method permits one to calculate a wave function out to large distances reliably and economically, a feature that makes the Ph-A method very useful.

For the repulsive case we focus on the description of the wave function in a repulsive barrier region, and analyze an example for a shape resonance in a Morse potential.

8.3 The Oscillatory Case

In this case the energy $E = k^2$ is larger that the potential V_T. The Schrödinger equation to be solved for a partial wave function ψ is

$$d^2\psi/dr^2 + k^2\psi = V_T\,\psi. \tag{8.3}$$

The total potential V_T is

$$V_T(r) = L(L+1)/r^2 + V(r), \tag{8.4}$$

where $V(r)$ is the atomic or nuclear potential (including the Coulomb potential) in units of inverse length squared, and L is the orbital angular momentum quantum number. Milne's non-linear equation for the amplitude $y(r)$ is given in this case by [2]

$$d^2y/dr^2 + k^2y = V_T\,y + k^2/y^3, \tag{8.5}$$

where the non-linearity is given by the last term in Eq. (8.5). In Eqs. (8.3)–(8.5) the factor $\hbar^2/2m$ has already been divided into the potential and into the energy, so that both are given in units of inverse length squared, and the wave number k is given in units of inverse length. The unit of length can be either fm for nuclear physics applications, or the Bohr radius [6] a_0 for atomic physics applications, but will not be explicitly indicated. The phase $\phi(r)$ is obtained from the amplitude y by [2]

$$\phi(r) = \phi(r_0) + k \int_{r_0}^{r} [y(r')]^{-2}\,dr', \tag{8.6}$$

but it can also be obtained without the knowledge of y [7]. The Eqs. (8.5) and (8.6) can be obtained by inserting Eq. (8.1) into (8.3), noting that the terms involving the phase $\phi(r)$ can be separated from the terms involving the amplitude $y(r)$, and setting to zero each term independently. An overall normalization is still arbitrary, but it can be fixed by demanding close agreement with the WKB approximation of a wave function. This is an important point since over the years, the WKB approximation has led to a much improved understanding of the solution of the Schrödinger equation. An excellent pedagogical description of the WKB approximation with several good references can be found in the book by Griffith [8]. The present phase-amplitude representation of the wave function is different from the one described by Calogero [3], in that

Eqs. (8.5) and (8.6) do not require the definition of auxiliary basis functions, as is the case for the Calogero method. The phase is obtained from the amplitude, while the reverse is the case for the Calogero method. Details are given in Appendix A of Ref. [9].

The Eq. (8.5) has been solved non-iteratively in the past by using some form of a finite difference computational method, such as one of Milne's predictor-corrector methods [10, 11], or by a Bulirsch–Stoer limit method [12], none of which we will use here. Instead, an iterative method will be described in the section below.

8.3.1   The Iterative Method of Seaton and Peach

The iterative method of Seaton and Peach [1] consists in rewriting Eq. (8.5) in the form

$$\frac{k^2}{y^4} = w + \frac{1}{y}\frac{d^2y}{dr^2},\tag{8.7}$$

where

$$w(r) = k^2 - V_T(r),\tag{8.8}$$

and calculating the solution of Eq. (8.7) by means of the iteration [1]

$$\frac{k}{y_{n+1}^2} = \left[w + \frac{1}{y_n}\frac{d^2y_n}{dr^2}\right]^{1/2}, \quad n = 0, 1, 2, \ldots \tag{8.9}$$

Here n denotes the order of the iteration. The initial value of y is given by the WKB approximation [4, 5]

$$\frac{k}{y_0^2} = w^{1/2}.\tag{8.10}$$

The advantage of formulating the iteration according to Eq. (8.9) is that y varies slowly with r, automatically and adiabatically approaching unity at large distances, and hence $(1/y_n)d^2y_n/dr^2$ is small compared to w. Near the origin of r this term may become large, and the iterations may not converge. In that case the solution of Eq. (8.9) should be started at a point sufficiently far from the origin, in a region where $[(1/y_n)d^2y_n/dr^2]$ is sufficiently small compared to w, depending on the accuracy desired, as will be described further below. A feature of the present Ph-A method is that no initial conditions are required to be imposed. Since the potential V_T in Eq. (8.8) approaches zero asymptotically, the amplitude y automatically approaches unity, asymptotically. This last property is very important, since the wave function derived from the present Ph-A method automatically also has unit amplitude asymptotically, even though the calculation is not required to be performed to asymptotic distances. This property does not hold for finite element or finite difference calculations. Equation (8.6) combined with the first order result is equivalent to the WKB approximation.

In summary, the iteration scheme (8.9) provides a method to iteratively improve the WKB approximation, since after convergence the resulting wave function is in much better agreement with benchmark wave functions than the WKB result for the numerical cases studied below.

8.3.2 The Spectral Computational Method

The spectral computational method consists in expanding the function y in the whole radial interval into a series of $N + 1$ Chebyshev polynomials $T_s(x)$, $s = 0, 1, 2, \ldots, N$,

$$y(x) = \sum_{s=1}^{N+1} a_s T_{s-1}(x), \quad -1 \le x \le 1. \tag{8.11}$$

The corresponding $N + 1$ support points $\xi_1, \xi_2, \ldots, \xi_{N+1}$ are the zeros of the polynomial T_{N+1}. Since the Chebyshev polynomials are defined in the interval $-1 \le x \le 1$ the quantities defined in the radial interval $0 \le r \le R_{\max}$ are mapped into the $x-$variable by a linear transformation. The expansion cutoff value $N + 1$ is set arbitrarily, but once chosen, the location and number of support points on the x-axis (and correspondingly on the r-axis) is determined. In the iterative algorithm that follows, only the values of the function y evaluated at the $N + 1$ support points are required, hence y and all the terms in Eq. (8.9) become vectors of dimension $N + 1$. Extensive use is made of the Clenshaw–Curtis matrix method (CC) [13] that relates the values of a function evaluated at the $N + 1$ mesh points to the expansion coefficients a_s of that function, and vice-versa by a simple known matrix [14] relation, as is explained in Chap. 5, Eq. (3.21), and also in Eq. (5.15).

The quantity $(d^2 y_n/dr^2)/y_n$ occurring in Eq. (8.9) is denoted as D_n

$$D_n = \frac{1}{y_n} \frac{d^2 y_n}{dr^2}. \tag{8.12}$$

If the second order derivative were obtained by replacing the T_s in Eq. (8.11) by their respective second derivatives,

$$d^2 y/dr^2 = \sum_{s=1}^{N+1} a_s (d^2 T_{s-1}(x)/dx^2)(dx/dr)^2,$$

one would run into numerical difficulties because the values of $d^2 T_s(x)/dx^2$ increase rapidly with the index s, and they would overwhelm the decrease of a_s with s. This is illustrated in Fig. 8.1.

Fig. 8.1 The values of dT_n/dx and d^2T_n/dx^2 for $x = -1$, as a function of the index n. For comparison, the values of n^2 (solid red line) and n^4 (dashed green line) are also displayed

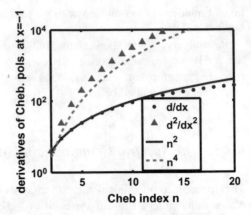

For example, for $s = 16$ and $x = -1$, $d^2T_s(x)/dx^2 = 2 \times 10^4$. Thus the value of N would have to be kept very small to around a value of approximately 10, and accuracy would be lost in calculating the phase, Eq. (8.6).

Instead, the values of D_n are recursively obtained as follows, By using $y_{n+1} = k^{1/2}(w + D_n)^{(-1/4)}$, one obtains after some algebra

$$D_{n+1} = \frac{5}{16}\frac{(w' + D_n')^2}{(w + D_n)^2} - \frac{1}{4}\frac{w'' + D_n''}{w + D_n}, \qquad (8.13)$$

where the primes mean derivatives with respect to r. By expanding D_{n+1} into a series of Chebyshev polynomials, and by differentiation of the latter (term by term), one obtains the derivatives of D_{n+1}. Hence one can proceed with Eq. (8.13), and from there one obtains the amplitude y_{n+1} according to Eq. (8.9). The errors mentioned above still do occur, but they are expected to be small compared to the derivatives of w. Alternatively, by dropping all the derivatives of D_n in Eq. (8.13), one obtains

$$D_{n+1} = \frac{5}{16}\frac{(V')^2}{(w + D_n)^2} + \frac{1}{4}\frac{V''}{(w + D_n)}, \qquad n = 0, 1, 2, \ldots \qquad (8.14)$$

Because Eq. (8.14) lacks the derivatives of D_n, it is to be used only at large distances where $D_n \ll w$. However, inclusion of the derivative improves the accuracy, as is shown in numerical examples given below.

8.3.3   A Non-iterative Analytical Solution of Eq. (8.14)

Provided that the iterations (8.14) converge, they can be replaced by an analytical method based on the solution of a cubic equation. The method is as follows. By adding to both sides of Eq. (8.14) the function w, and by defining

$$z_n = w + D_n,$$ (8.15)

Eq. (8.14) is equivalent to

$$z_{n+1} = \frac{(5/16)(V')^2}{z_n^2} + \frac{(1/4)V''}{z_n} + w, \quad n = 0, 1, 2, \ldots.$$ (8.16)

If the iterations converge, then for a very large value n_∞ of n the values of the $z's$ will no longer change with n. In other words, we have

$$z_{n_\infty+1} \simeq z_{n_\infty} = z.$$ (8.17)

Replacing in Eq. (8.16) all the z_n's by z, and after some algebraic rearrangement one obtains the cubic equation

$$z^3 = A + Bz + w z^2,$$ (8.18)

with

$$A(r) = (5/16)(V')^2 \text{ and } B(r) = (1/4)V''.$$ (8.19)

This equation can be solved by standard methods as follows [15]. By defining $p = -(B + w^2/3)$, $q = -(2w^3 + 9wB + 27A)/27$, and

$$J = -4p^3 - 27q^2,$$ (8.20)

the three roots of Eq. (8.18) can be expressed in terms of p, q, w, and $(-3J)^{1/2}$. If $J \leq 0$, so that $(-3J)^{1/2}$ is real, then one of the roots $z^{(1)}$ is real, and the other two are complex conjugates of each other. Here

$$z^{(1)} = (\mathscr{A} + \mathscr{B} + w)/3,$$ (8.21)

where

$$\mathscr{A} = -(27/2)q + (3/2)(-3J)^{1/2},$$ (8.22)

and

$$\mathscr{B} = -3p/\mathscr{A}.$$ (8.23)

In the numerical examples given below, it is found that after 5 iterations, the value of $\omega + D_5$ differs from $z^{(1)}$ by less than 10^{-11}, showing that the iterations of Eq. (8.14) converge. In Eqs. (8.14) and (8.16) derivatives of the potential with respect to r are required. If the potential V is given by analytic expressions, as is the case for the numerical examples described below, the derivatives can be calculated analytically. Otherwise the derivatives have to be calculated by some type of numerical interpolation procedure. If the iterations are performed according to Eq. (8.13), then the above formalism still applies, with V' replaced by $-(w' + D_n')$ and V'' replaced by $-(w'' + D_n'')$.

8.3.4 The Algorithm

1. For the first iteration, D_0 is set to zero in Eq. (8.9), and the values of the square root in Eq. (8.9) are evaluated at the support points $\{\xi\}$. The values of y_0 are then also obtained at the support points. The non-linearity causes no problems because these quantities can be manipulated numerically (square roots, inverses, etc.) support point by support point, without the Chebyshev expansion being invoked.

2. Subsequently the Chebyshev expansion coefficients a_n of y_0 are obtained by means of the (CC) method, and in terms of those the integral in Eq. (8.6) can be carried out by means of the Gauss–Chebyshev method [14, 16]. The result is transformed back into the support points via the (CC) method, and the wave function defined via Eq. (8.1) is obtained at the support points. The result is the WKB approximation.

3. In order to obtain the next order function y_1, the required second derivative of y_0 is obtained via Eq. (8.14), with D_n replaced by D_0.

4. If iterations of Eq. (8.9) are performed, then D_{n+1} is obtained in terms or D_n from Eq. (8.14), with $n = 1, 2, \ldots$, else the cubic equation (8.18) is solved analytically. The Chebyshev expansion of $1/y_n^2$, Eq. (8.11) is invoked only in order to calculate the phase given by Eq. (8.6) at the support points, and to subsequently interpolate the results to an equidistant mesh, as required for graphical purposes.

5. If a comparison is required of the Ph-A wave function with a wave function ψ_C calculated by some other means (the "C" stands for "numerically calculated"), for instance in a short radial interval $[0, r_0]$, then in addition to the steps 1–4, the procedure described in Sect. 8.3.5 is required to renormalize ψ_C and to obtain the Ph-A starting phase ϕ_0 at r_0.

In the present numerical examples 5 iterations sufficed in order for the difference between y_{n+1} and y_n to be less than 10^{-8} for all support points in the whole radial interval. The difference between y_5 and the non-iterative solution of Eq. (8.18) was also found to be less that 10^{-8}, which is an indication that the iterations converged for these cases. The integral in Eq. (8.6) required to calculate the phase ϕ is performed by a Gauss–Chebyshev method [14, 16] that is well suited to this type of spectral expansion since it only requires the values of the expansion coefficients a_s. Situations that involve imaginary local wave numbers and the respective turning points, as is the case in the presence of repulsive barriers or where the potential is everywhere repulsive, are described in the Sect. 8.4 below.

8.3.5 Connection with a Conventional Wave Function

It is common that at small distances the potentials have a repulsive core, and the centripetal potential also introduces a strong repulsive part. For these cases, and also in the presence of repulsive barriers, it is preferable to calculate the wave function by conventional numerical means from the origin to a point r_0, and connect this wave function to the Ph-A wave function at r_0, so that the wave function can be propagated out to large distances by the phase-amplitude method. Point r_0 should be located to the right of the left most turning point meaning that it should be contained in the oscillatory region of the wave function. A procedure for implementing this connection will be described further on. This method does not use the conventional matching procedure based on the Wronskian of the wave functions.

If the wave function obtained by some numerical algorithm for $0 \leq r \leq r_0$ is denoted as ψ_C (C for "computational"), and if that wave function is to be extended beyond r_0 by means of the Ph-A representation for which ϕ_0 and y_0 are the initial phase and amplitude at point r_0, then two quantities need to be obtained from ψ_C, namely ϕ_0 and the normalization factor $\varkappa$ of ψ_C so that

$$\varkappa \psi_C = \psi = y \sin(\phi) \text{ for } r \simeq r_0. \tag{8.24}$$

The wave function defined in Eq. (8.1) has the property that at large distances the amplitude goes to unity. But for long range potentials it is not possible to assure that the conventional wave function $\varkappa \psi_C$ satisfies this requirement, unless $\varkappa$ is obtained by a Wronskian matching procedure involving basis functions g and f that are known to approach unity at large distances. Such basis functions are known analytically for the Coulomb case, and for short ranged potentials, but are difficult to obtain for general long range potentials. The present Ph-A procedure avoids the need for such basis functions.

In the expansion of a wave function over the interval $[r_0, r_{\max}]$ in terms of Chebyshev polynomials of the first kind, [17] the support points $\{\xi\}$ do not include r_0 but have values $r_0 < \xi_1 < \xi_2 \cdots < \xi_{N+1} < r_{\max}$ that are described below. The corresponding values of the amplitude at the support points are denoted as $y_1, y_2, \ldots,$ and the respective phases are given by

$$\phi_1 = \phi_0 + \varphi_1, \ \phi_2 = \phi_0 + \varphi_2, \ldots, \tag{8.25}$$

where

$$\varphi_i = k \int_{r_0}^{r_i} [y(r')]^{-2} dr', \ i = 1, 2, \ldots, N + 1. \tag{8.26}$$

The values of y_i and φ_i, $i = 1, 2, \ldots$ are known from the numerical procedure, either from Eq. (8.9), or from Eq. (8.18). By defining

$$\Delta = \varphi_2 - \varphi_1, \tag{8.27}$$

and

$$\mathscr{O} = \frac{\psi_C(r_2)/y_2}{\psi_C(r_1)/y_1},$$ (8.28)

and after making use of some trigonometric identities, one obtains [18]

$$\frac{\mathscr{O} - \cos(\Delta)}{\sin(\Delta)} = \cot(\phi_0 + \varphi_1).$$ (8.29)

The value of ϕ_0 can be obtained form Eq. (8.29) and $\varkappa$ is given by

$$\varkappa = \frac{y_1 \sin(\phi_0 + \varphi_1)}{\psi_C(r_1)}.$$ (8.30)

8.3.6 Numerical Results for the $1/r^3$ Case

The feasibility of the present approach will be demonstrated by means of two examples, for which the potential V_T is everywhere attractive. In these examples the potential has a long range tail proportional to either r^{-3} or r^{-1}, respectively. In order to mitigate the singularities at the origin of r, a "rounding" procedure is introduced that replaces r by a constant in the limit $r \to 0$ by means of an analytic transformation from r to $\mathscr{R}$

$$\mathscr{R}(r) = r/[1 - \exp(-r/t)],$$ (8.31)

but does not alter the potential at distances much larger than the rounding parameter t. Hence $V(r)$ is changed to $V[\mathscr{R}(r)]$.

In the $1/r^3$ case the potential is given by

$$V_3(r) = -1.6224 \times 10^4/\mathscr{R}^3,$$ (8.32)

where the $\mathscr{R}(r)$ is given by Eq. (8.31) in terms of r. The purpose of the "rounding" is to permit convergence of the iterations near the origin, which would not be the case for a potential that diverges near the origin.

The value of this potential as illustrated in Fig. 8.2 is appropriate for atomic physics applications, when the two interacting molecules have dipole moments. The reason this $1/r^3$ long range nature was chosen is because this case did not get addressed successfully by means of an iterative Born-approximation method [19], while it is well described with the Ph-A method. At $r = 2500$ the value of V is $\simeq 10^{-6}$, while near $r = 0$ its value is 1900. The latter is unphysically large, but the example serves to illustrate the numerical method described here. The corresponding wave function is highly oscillatory at small distances, with an amplitude and wave length that varies substantially with distance, as is illustrated in Fig. 8.3, yet it is well reproduced by the Ph-A method, as described further on.

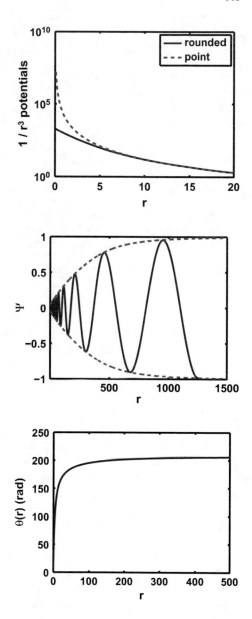

Fig. 8.2 The "rounded" $1/r^3$ potential, given by Eq. (8.32) with the rounding parameter $t = 2$. The units are in inverse length squared, since the potential, in energy units, has been multiplied by the factor $2m/\hbar^2$. The dashed line indicates the non-rounded potential, with $\mathcal{R}$ replaced by r

Fig. 8.3 The wave function calculated by the S-IEM method, corresponding to potential (8.32), with $t = 10$ in Eq. (8.31) for $k = 0.01$. Asymptotically its amplitude is unity. The dashed lines top and bottom represent the amplitude y and $-y$, respectively, calculated from Eq. (8.9) for the same potential

Fig. 8.4 The phase shift, defined in (8.33) for potential (8.32) with $t = 2$ for $k = 0.01$. Beyond $r = 500$ the phase shift still changes because the potential is not yet constant, as further explained in the text

The corresponding phase shift $\theta(r)$, defined as

$$\theta(r) = \phi(r) - (kr - L\pi/2), \tag{8.33}$$

is illustrated in Fig. 8.4. Because of the long range nature of the potential, the phase shift still changes for $1000 < r < 2000$ by $\simeq 0.3$ rad. Their values are 207.45057 rad

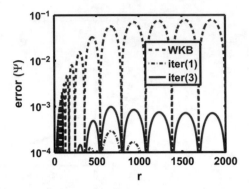

Fig. 8.5 Error of the PhA wave function for the "rounded" $1/r^3$ potential (8.32), with the rounding parameter $t = 2$, and $k = 0.01$. The Ph-A calculation uses 201 Chebyshev expansion functions in the radial domain [0, 2000]. The order n of the iteration is shown in parenthesis in the legend. The result for $n = 1$ is shown by means of the blue dotted line at the bottom of the graph. For $n > 3$ the result does not change perceptively any further, i.e., convergence sets in

and 207.74413 rad for $r = 1000$ and 2000, respectively. The agreement of the Ph-A wave function with a benchmark wave function denoted as S-IEM, calculated by a spectral method [14] and normalized according to Sect. 8.3.5, is shown in Fig. 8.5.

Noteworthy is the fact that in this example only 201 expansion terms in Eq. (8.11) have been used to calculate the amplitude and phase for the whole radial interval [0, 2000].

An evaluation of the error of the Ph-A wave function is obtained by plotting the absolute value of the difference of the Ph-A and the S-IEM wave functions. The result for the cases $k = 0.01$ as calculated in the radial interval [0, 2000] with 201 Chebyshev expansion functions is illustrated in Fig. 8.5.

The iterative and non-iterative solutions converge towards each other. This is shown by the fact that the difference between 5th iterative solution of Eq. (8.14) and the cubic solution of Eq. (8.18) is $\leq 10^{-12}$.

A calculation in a smaller radial interval, starting at $r_0 = 30$, and ending at $r = 200$, using only 51 Chebyshev functions, is illustrated in Fig. 8.6. The initial value of the phase ϕ_0 was obtained according to Sect. 8.3.5. The basic conclusion is that the Ph-A method provides an excellent improvement over the WKB approximation. Please note that in both Figs. 8.5 and 8.6, the results for the first iteration are more accurate than for the third iteration. However convergence sets in as the number of iterations increases, as is shown in Fig. 5 of Ref. [9].

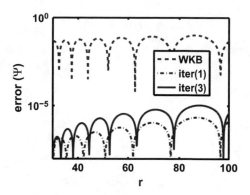

Fig. 8.6 Same as Fig. 8.5 in the radial interval [30, 200], using 51 Chebyshev functions. The equations in Sect. 8.4 are used to start the Ph-A wave function at $r = 30$. Please note the change in scale from Fig. 8.5

8.3.7 Numerical Results for the 1/r Coulomb Potential Case

The conventional equation for the wave function $\chi(\rho)$ describing the Coulomb case is

$$\frac{d^2\chi}{d\rho^2} + \left[1 - \frac{2\eta}{\rho} - \frac{L(L+1)}{\rho^2}\right]\chi = 0, \tag{8.34}$$

where ρ describes the radial distance, and η describes the strength of the Coulomb interaction. The corresponding potential in Eq. (8.3) is given by

$$V(r) = \frac{Ze^2}{r}\frac{2m}{\hbar^2} = \frac{\bar{Z}}{r}, \tag{8.35}$$

where $\bar{Z}$ has units of inverse length and is related to 2η according to

$$\bar{Z} = 2\eta k, \tag{8.36}$$

where

$$2\eta = Z\frac{e^2}{\hbar c}\left(\frac{mc^2}{E}\right)^{1/2}, \tag{8.37}$$

and where the equation for the "rounded" Coulomb potential is

$$V(r) = \frac{Ze^2}{\mathcal{R}}\frac{2m}{\hbar^2} = \frac{\bar{Z}}{\mathcal{R}}. \tag{8.38}$$

In the earlier equations, m is the mass, E the energy of the incident particle, Z the number of protons in the nucleus, e the charge of a proton, $\hbar$ is Planck's constant, and $\rho = kr$. For the attractive case, Z and η are both negative. Thus by setting $k = 1$ and $\bar{Z} = 2\eta$ in Eq. (8.3), one obtains a solution to Eq. (8.34). Asymptotically the Coulomb wave function approaches

Fig. 8.7 The phase θ as a function of distance for the Coulomb case described in the text

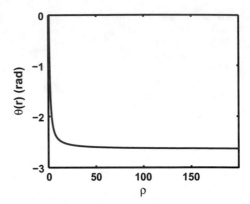

$$\chi(\rho) \to \sin(\rho - \eta \ln(2\rho) - L\pi/2 + \theta_\infty), \tag{8.39}$$

where θ_∞ is the phase shift, whose value depends on how the potential differs from the point Coulomb potential. The position dependent phase is defined as

$$\theta(\rho) = \phi(\rho) - [\rho - \eta \ln(2\rho) - L\pi/2], \tag{8.40}$$

which asymptotically approaches the phase shift θ_∞.

An example for the Ph-A wave function for a "rounded" Coulomb potential with the rounding parameter $t = 2$, and $\eta = -2$, and started at point $\rho = 30$ has been calculated. Good agreement with the S-IEM wave function was obtained, as shown in Figs. 8.8 and 8.10. The required phase ϕ_0 was obtained according to the procedure described in Sect. 8.3.5 by making use of the S-IEM wave function calculated in the interval [0, 32]. By means of this procedure the S-IEM wave function is also normalized, so that asymptotically its amplitude tends toward unity, should it be calculated numerically out to large distances.

The phase, given by Eq. (8.40), and illustrated in Fig. 8.7, still changes in the second significant figure for $\rho \simeq 100$, and it stabilizes in the fourth significant figure to -2.6399 rad beyond $\rho \simeq 2500$. That level of stabilization is compatible with the fact that at these distances the rounded Coulomb potential differs from the point Coulomb potential by 4×10^{-2} and 2×10^{-3} respectively, and it demonstrates that the long range $1/\rho$ character is taken into account reliably.

The accuracy of the Ph-A wave function for $\eta = -1$ is illustrated in Fig. 8.8. The first iteration gives an accuracy that is 10^4 times better than the WKB accuracy, and subsequent iterations (they do converge) make only an additional slight improvement.

It is worth noting that the accuracy is practically independent of position for $\rho > 500$. This is a property of the Chebyshev expansion, for which the error distributes itself uniformly across the whole radial domain.

For the case of a point Coulomb potential the iterative method does not converge for small values of ρ. Although the solution of Eq. (8.18) was obtained non-iteratively, it gave an error for the point Coulomb wave function of more than 10%. This can

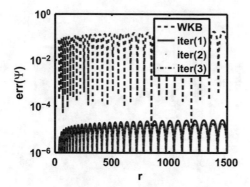

Fig. 8.8 The accuracy of the Ph-A wave function for the rounded Coulomb potential, with rounding parameter $t = 2$ and for $\eta = -1$. The radial interval is [30, 1500], and the number of Chebyshev polynomials used in this interval is 51. Please note that the improvement of the accuracy over the WKB approximation is dramatic. Further, the error is uniform given that it does not change with position r

Fig. 8.9 The functions D/w for the point and rounded Coulomb potentials for $\eta = -2$

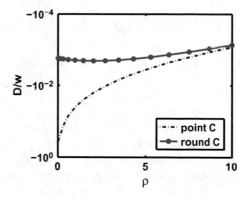

be understood by examining the value of D/ω, illustrated in Fig. 8.9. This figure shows that the value of D/w for $\rho < 1$ is close to unity for the point Coulomb case while for the rounded Coulomb case it is close to 10^{-3}. Hence for the point Coulomb case the solution of Eq. (8.18) should not be expected to be accurate. However, in the rounded Coulomb case, it should be expected to be accurate to at least 10^{-3}. The numerical calculations show that the accuracy is of the order of 10^{-5}, as illustrated in Fig. 8.10, but if dD/dr is included to first order in Eq. (8.18) then the accuracy increases by an order of magnitude more to less than 10^{-6}. The symbol ∞ in the legend of Fig. 8.10 indicates that the solution of the cubic equation was used in the calculations.

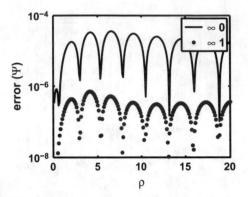

Fig. 8.10 Accuracy of the Ph-A wave function for the rounded Coulomb potential with $\eta = -2$ in the radial interval $0 < \rho < 200$, using 101 Chebyshev polynomials. The results marked as $\infty 0$ and $\infty 1$ are obtained from the solution of the cubic equations, respectively, without, and with the inclusion of the first order derivative of D in Eq. (8.16). The message of this graph is to show that the iteration via method 2 works as well as via method 1

8.3.8   Numerical Details

The calculations are done with MATLAB on a desk PC using an Intel TM2 Quad, with a CPU Q 9950, a frequency of 2.83 GHz, and a RAM of 8 GB. The calculation uses typically between $N + 1 = 51$ and 201 Chebyshev expansion polynomials. The computing time in the radial interval [30, 1500] and the corresponding accuracies for a Coulomb wave function calculation are given in Table 8.1. The first column lists the number minus one of Chebyshev polynomials used in the calculations. Column 2 gives the computing time for the non-iterative solution of Eq. (8.18). Column 3 gives the computing time for carrying out five iterations of Eq. (8.14), and column 4 gives the resulting accuracy. The computing time for the Ph-A iterations depends weakly on the number of Chebyshev functions $N + 1$, regardless of the size of the radial interval, and depends weakly on the value of k. By contrast, the S-IEM calculation in the radial interval [0, 1500] for the whole wave function depends on the value of k, requiring 0.20 s for $k = 0.01$ and 0.29 s for $k = 0.1$. The times given in Table 8.1 do not include the time to interpolate the y and ϕ to a fine equidistant radial mesh. Interpolating to an equispaced radial mesh size of step length $h = 0.1$ depends on the

Table 8.1 Ph-A computation times, and Ph-A wave function accuracy

N	Cubic time (s)	5 iter time (s)	Accuracy
16	0.045	0.89	10^{-3}
25	0.048	0.90	10^{-4}
50	0.060	0.90	10^{-5}
100	0.061	0.91	10^{-5}

size of the radial interval. For the radial interval $[0, 40]$ the fine mesh interpolation requires $0.8\,\text{s}$, and for the radial interval $[40, 2000]$ the interpolation takes between 170 and $180\,\text{s}$.

8.4 The Exponential Case

It is the purpose of the present section to adapt the Ph-A method to the radial region where $E < V$, that is the region forbidden classically. However, the restrictions for the validity of the WKB approximation are valid also for the method presented here, because the method is accurate only in a region sufficiently far away from the turning points such that the variations of the potential at a distance of the local wave length are small compared to the potential itself [8]. The non-linear method of Milne will still apply, and the iteration scheme of Seaton and Peach with the suitable modifications, will also apply. The main difference from the case that $E > V$, described in Sect. 8.3, is that the wave function now increases or decreases exponentially, and near the turning points where $E = V$, the iterations do not converge well or not at all. The assumption here is that the asymptotic energy $E = k^2$ is still positive, meaning that the asymptotic wave number k is real.

8.4.1 The Formalism

The Schrödinger equation to be solved is Eq. (8.3). In the barrier region, where $V > E$, the local wave number κ is given by

$$\kappa(r) = \sqrt{\tilde{w}(r)}, \qquad (8.41)$$

where

$$\tilde{w}(r) = V(r) - k^2 > 0. \qquad (8.42)$$

If one inserts into Eq. (8.3) the ansatz

$$\psi^{(-)}(r) = y^{(-)}(r) \exp(-\Phi^{(-)}(r)) \qquad (8.43)$$

and

$$\psi^{(+)}(r) = y^{(+)}(r) \exp(\Phi^{(+)}(r)), \qquad (8.44)$$

where $\psi^{(-)}$ and $\psi^{(+)}$ satisfy each Eq. (8.3), then the full wave function is given by

$$\psi(r) = A\psi^{(-)}(r) + B\psi^{(+)}(r) \qquad (8.45)$$

and one finds that

$$y^{(+)} = y^{(-)} = \tilde{y} \tag{8.46}$$

and

$$\Phi^{(+)} = \Phi^{(-)} = \Phi(r). \tag{8.47}$$

In the above the common amplitude $\tilde{y}$ obeys the equation

$$\frac{d^2\tilde{y}}{dr^2} - \tilde{w}\,\tilde{y} = -\frac{k^2}{\tilde{y}^3} \tag{8.48}$$

and the common phase Φ is given by the simple quadrature

$$\Phi(r) = \int_a^r \frac{k}{\tilde{y}^2(r')} dr'. \tag{8.49}$$

A strong formal similarity exists between Eqs. (8.7) and (8.48), however $d^2 y/dr^2$ is replaced by $-d^2\tilde{y}/dr^2$, which mathematically makes a big difference. The two turning points at the extremities of the barrier region are T_1 and T_2, and the region suitable for the Ph-A method is located in $[a, b]$, which is contained between T_1 and T_2. (Hence $T_1 < a \leq r \leq b < T_2$). According to Eq. (8.45) the complete Ph-A wave function in $[a, b]$ is given by

$$\psi(r) = \tilde{y}(r)\left[A\,e^{-\Phi(r)} + B\,e^{+\Phi(r)}\right]; \quad a \leq r \leq b, \tag{8.50}$$

where the coefficients A and B are determined by the connection formulae across the turning points.

The validity of Eqs. (8.46)–(8.50) can be verified by inserting Eq. (8.50) into (8.3), and setting to zero each of the terms that are multiplied respectively the factors A or B. Once the phase is defined according to Eq. (8.49), then the relationship between phase Φ and amplitude $\tilde{y}$ is determined uniquely, but other relationships are also possible [20], as is the case for the Calogero's Ph-A formalism, described in Appendix A of Ref. [9].

8.4.2   Iterative Solution

In this section the iterative method of Seaton and Peach [1] is extended to the case of the barrier region [21], where $\tilde{w} = V - k^2 > 0$. By re-writing Eq. (8.48) and taking square roots, one obtains

$$\frac{k}{\tilde{y}_{n+1}^2} = (\tilde{D}_n + \tilde{w})^{1/2}, \quad n = 0, 1, 2, \ldots \tag{8.51}$$

where $\tilde{w}$ is defined in Eq. (8.42), where

$$\tilde{D}_n = -\frac{d^2 \tilde{y}_n / dr^2}{\tilde{y}_n} \tag{8.52}$$

and where $\tilde{D}_0 = 0$. The resulting value of $\tilde{y}_1$ is identical to the WKB approximation

$$\tilde{y}_1 = \tilde{y}_{WKB} = (V/k^2 - 1)^{-1/4}, \quad a \le r \le b, \tag{8.53}$$

and hence the phase and the wave function (8.50) become identical to their WKB values. Similarly to Eq. (8.13), in order to avoid a numerical loss of accuracy when calculating the second order derivative of $\tilde{y}$, it is preferable to obtain $\tilde{D}_n$ by a recursion relation which in the present case takes the form

$$\tilde{D}_{n+1} = -\frac{5}{16} \frac{(\tilde{D}_n' + \tilde{w}')^2}{(\tilde{D}_n + \tilde{w})^2} + \frac{1}{4} \frac{(\tilde{D}_n'' + \tilde{w}'')}{(\tilde{D}_n + \tilde{w})}, \quad n = 0, 1, 2, \ldots, \tag{8.54}$$

where "primes" denote derivatives with respect to r. For $n = 0$, one has $\tilde{D}_0 = 0$, and hence all its derivatives are zero. The derivatives of $\tilde{w}$ are equal to the derivatives of V. They can be calculated analytically if the analytic expression for V is known, as is the case for the numerical examples described in the next section. The advantage of obtaining $\tilde{D}_{n+1}$ by means of Eq. (8.54) instead of calculating the second derivative of $\tilde{y}_{n+1}$ directly is that the quantity $\tilde{D}_{n+1}$ and its derivatives are small compared to $\tilde{w}$ and its derivatives, and hence the effect of the errors of the Chebyshev expansion of $\tilde{D}_{n+1}$ and its derivatives are reduced. A numerical comparison of the potential V and the quantity $\tilde{D}$ is presented in connection with an application of the Ph-A method to a Coulomb potential.

There are two methods for obtaining $\tilde{D}$. One consists in inserting the values of $\tilde{D}_{n-1}$ and its derivatives into Eq. (8.54) in order to obtain $\tilde{D}_n$. Next, the derivatives of $\tilde{D}_n$ calculated numerically are inserted into the right hand side of Eq. (8.54) and the iteration for $\tilde{D}_n$ continues until the iteration converges for n_{max}. By inserting the result into Eq. (8.51) one obtains the values of $\tilde{y}_{n_{max}}$ for all support points, hence the phase for all support points can be obtained from the quadrature indicated in Eq. (8.49) from which the functions $\psi^{(\pm)}$ can be obtained. This procedure is especially economical if a spectral expansion of all the functions in terms of Chebyshev polynomials is implemented, given that the integrals can be performed by means of the methods described in Chap. 5. The coefficients A and B required for the full wave function in Eq. (8.50) are obtained by a connection formula across the left turning point, as described below.

A second iterative method for obtaining $\tilde{D}$ is as follows: one assumes fixed values for the first and second derivatives of $\tilde{D}$, denoted as $\tilde{D}_M'$ and $\tilde{D}_M''$ and inserts them into the numerators of Eq. (8.54). This can be written as

$$\tilde{z}_{n+1} = -\frac{5}{16}\frac{(\tilde{D}'_M + \tilde{w}')^2}{(\tilde{z}_n)^2} + \frac{1}{4}\frac{(\tilde{D}''_M + \tilde{w}'')}{(\tilde{z}_n)} + \tilde{w}, \quad n = 0, 1, 2, \ldots, \qquad (8.55)$$

where

$$\tilde{z}_n = \tilde{D}_n + \tilde{w}. \qquad (8.56)$$

After convergence at $n = n_{max}$, the value of $\tilde{z}_{n_{max}+1}$ can be set equal to $\tilde{z}_{n_{max}} = \tilde{z}_M$, and Eq. (8.55) can be transformed into the cubic equation

$$\tilde{z}_M^3 = -\frac{5}{16}(\tilde{D}'_M + \tilde{w}')^2 + \frac{1}{4}(\tilde{D}''_M + \tilde{w}'')(\tilde{z}_M) + \tilde{w}(\tilde{z}_M)^2. \qquad (8.57)$$

The solution of Eq. (8.57) does not require iterations but can be solved by standard algebraic means [15], as explicitly described after Eq. (8.18). The derivatives of the resulting values of $\tilde{z}_M$ are calculated numerically and are denoted by $\tilde{z}'_{M+1}$ and $\tilde{z}''_{M+1}$. They are inserted into Eq. (8.57) and the next value of $\tilde{z}$, denoted a $\tilde{z}_{M+1}$ is obtained, and so on. The values of $\tilde{y}$ are obtained from Eq. (8.51), $\tilde{y}_M = (\tilde{z}_M/k^2)^{-1/4}$, and the calculation for the phase and wave function proceeds as described above. In the numerical examples given in Sect. 8.3.6, the iteration index M is denoted as n. It is shown in Ref. [21] that when the iterations with method 1 converge (which is in the region of validity of the WKB approximation), then the iterations with method 2 will also converge and give the same result. But the reverse is not always the case. In this case the results of method 2 are not to be trusted.

The numerical calculation consists in expanding the (still unknown) function $\tilde{y}(r)$ in the radial domain $[a \leq r \leq b]$ in terms of Chebyshev polynomials $T_n(x)$, as described in Eq. (8.11), $\tilde{y}(x) = \sum_{s=1}^{N+1} \tilde{a}_s T_{s-1}(x)$, $-1 \leq x \leq 1$. Once the coefficients $\tilde{a}_n$ are known, Eq. (8.11) can be used to evaluate $\tilde{y}$ at any continuous point in $[a, b]$. Because both $\tilde{y}$ and Φ vary more slowly with distance than the wave functions themselves the number of support points can be quite small, as is demonstrated in the numerical examples presented in Sect. 8.3.6.

8.4.3   Connection Formulae

The wave function calculated numerically to the left side of the left turning point by some means other than the Ph-A method is denoted as $\psi_C(r)$ (the subscript C denotes "Conventional Computation"). This wave function is continued numerically through the left turning point T_1 to two points r_1 and r_2 located in the barrier region where the Ph-A method is reliable, i.e., at $T_1 < a < r_1 < r_2 << b < T_2$. At these two points the Ph-A wave function is evaluated,

$$\psi_{Ph-A}(r_i) = \tilde{y}(r_i)[Ae^{-\Phi(r_i)} + Be^{\Phi(r_i)}], \quad i = 1, 2. \qquad (8.58)$$

in which the coefficients A and B still to be determined, but the values of $\tilde{y}$ and Φ are known from the solution of the Ph-A equations. By equating the values of $\psi_{Ph-A}(r_i)$ to the values of $\psi_C(r_i)$, $i = 1, 2$, one obtains the matrix equation

$$
\begin{pmatrix} \tilde{y}_1 e^{-\Phi_1} & \tilde{y}_1 e^{+\Phi_1} \\ \tilde{y}_2 e^{-\Phi_2} & \tilde{y}_2 e^{+\Phi_2} \end{pmatrix} \begin{pmatrix} A \\ B \end{pmatrix} = \begin{pmatrix} \psi_C(r_1) \\ \psi_C(r_2) \end{pmatrix}
\tag{8.59}
$$

that can be solved for the coefficients A and B. In the above $\tilde{y}_i = \tilde{y}(r_i)$ and $\Phi_i = \Phi(r_i)$, with $i = 1, 2$. In order to obtain reliable values for the coefficients A and B the condition number of the matrix in Eq. (8.59) has to be sufficiently small. However, at this stage the functions ψ_C have an arbitrary normalization. That normalization is transmitted to the values of A and B. Once the asymptotic value of the wave function is known and normalized appropriately, then the whole wave function can be normalized accordingly.

In order to connect the resulting Ph-A wave function across the right turning point T_2 to a conventional numerical wave function ψ_C, the methods described in Sect. 8.2 can be adapted to the present situation. Two independent solutions of the Schrödinger equation can be calculated in the region across the right-hand turning point, and the appropriate linear combination can be obtained by the method described above. A turning point is the location where the energy E and the potential V are equal. The radial region where $E < V$ is classically forbidden, nevertheless the quantum-mechanical wave function penetrates this region.

8.4.4 Numerical Example for the Morse-Like Potential

Shape resonances play an important role in many areas of physics. A case for atomic physics is described in Ref. [22]. The Morse potential described in Ref. [23] is again chosen for the present application, because the two resonances that occur for this potential are known and well studied [23]. The accuracy analyses described in that study and in Ref. [16] were possible, because the phase shifts are known analytically for the Morse potential case. It was shown in Fig. 2 of Ref. [16] that using a sixth order Numerov method for the solution of the Schrödinger equation the results were six orders of magnitude *less* accurate than that of a solution based on a spectral integral equation method denoted as S-IEM [24]. The latter method will also be used in the present investigation as a benchmark comparison result. The main point made in the present section is to show that the Ph-A method is capable of generating both the exponentially increasing and the decreasing solution with equal accuracy, given that the phase and the amplitude for each are the same.

Two cases will be examined for the incident momentum k: (1) In a region of resonance, and (2) In a region far from a resonance. The resonance region presents a challenging test of accuracy, because some of the wave functions do decrease with distance r in that region, while the errors of the computational method introduce increasing contributions in the barrier region.

Fig. 8.11 The Morse potential is illustrated by the curved blue line. The horizontal lines indicate the location of the resonance energies, $E_1 = 1.1779$ (green solid line) and $E_2 = 5.57$ (red dashed line), all in units of inverse length squared

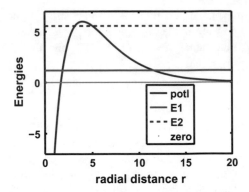

The potential $V(r)$ is given in Chapter 6 [23], and repeated here for convenience, by

$$V(r) = V_0\, e^{-(r-r_e)\,\alpha} \left[2 - e^{-(r-r_e)\,\alpha}\right], \tag{8.60}$$

with

$$V_0 = 6;\, r_e = 4;\, \alpha = 0.3. \tag{8.61}$$

Here V_0 is in units of inverse length squared r_e in units of length, and α in units of inverse length. The difference from Chap. 6 is that here $V_0 = 6$. This potential illustrated in Fig. 8.11 has a finite attractive valley near the origin, and a repulsive region beyond $r \simeq 2$. It supports one bound state and two resonances at energies E_1 and E_2, both indicated in the figure. The wave functions in the two resonance regions are shown in Figs. 8.12, 8.13 and 8.14. The numbers in the legends label the momentum values k that span the respective resonances, as indicated in the captions to these figures.

The wave functions are obtained from the numerical S-IEM [24] solution of the wave equation in the radial region ($0 \le r \le 100$), and they are normalized such that near $r = 100$ the oscillation maxima are set close to unity.

8.4.4.1   Resonance 1

The two turning points for the energy $E_1 = 1.1779$ in the vicinity of this resonance occur near $T_1 \simeq 1.9$ and $T_2 \simeq 11.7$. The region of convergence of the Ph-A approximation occurs in the radial interval that is approximately given by $r \in [3.5, 5.5]$. The values of the Ph-A amplitude $\tilde{y}$ are obtained by the iterative method given by Eq. (8.57), with the initial value of the derivatives of $\tilde{D}$ given by the WKB approximation. The rate of convergence of the iteration is shown in Fig. 8.15.

This figure shows that the error of y for the WKB approximation is between 10^{-2} and 10^{-3} for the radial interval shown, while the error after the 8th iteration is of the order of 10^{-5}.

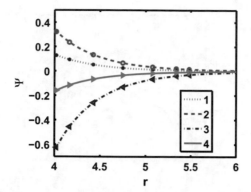

Fig. 8.12 The continuous lines illustrate the wave function results in the barrier region obtained by solving the Schrödinger equation for a Morse potential using the S-IEM method for $0 \leq r \leq 100$, for the various energies given by k^2. Here the wave numbers k spanning the resonance region #1 are given by $k = 1.08526787 + (n - 1) \times 10^{-8}$, with $n = 1, \ldots, 4$. The barrier region extends from $r = 2$ to $r = 12$. The discrete symbols represent the results of an independent Phase-Amplitude calculation described in the text. The good agreement of the symbols with the curves is an indication of the validity of the Ph-A method in the resonance region

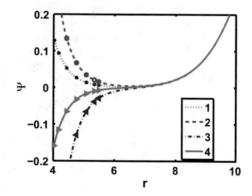

Fig. 8.13 The numerical solutions of the Schrödinger equation are the same as those in Fig. 8.12. In the present figure the radial region extends beyond the barrier for the energies indicated in Fig. 8.12. The figure shows that these functions become almost identical at large distances, even though in the region near the origin they are quite different

The phase $\Phi(r)$ is displayed in Fig. 8.16.

The amplitude $\tilde{y}$ for $k = 1.0855$ is illustrated in Fig. 8.17 and the wave functions $\psi^{(+)}$ and $\psi^{(-)}$, defined as $\tilde{y} \exp(\pm\Phi)$, are illustrated in Fig. 8.18.

The calculation of the Ph-A wave function defined in Eq. (8.50) requires the calculation of the coefficients A and B, which according to Eq. (8.59) depend on the wave function ψ_C obtained at two (arbitrary) points r_1 and r_2 located inside the region $[a, b]$ near the left end. In the present case ψ_C is calculated by means of the S-IEM method starting from the origin, and normalized asymptotically to have

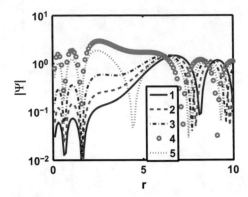

Fig. 8.14 Resonance 2 occurs near $k = 2.24$. The wave functions in that region have k-values given by $k = 2.24 + (n - 1) \times 0.04$, with $n = 1, \ldots, 5$. The barrier region extends from $r = 3$ to $r = 6$, and is too small for either the WKB or the present calculation to give meaningful results. The figure shows that for $k = 2.24$ and 2.28 ($n = 1$ and 2) the wave function inside the barrier region (blue and green dashed lines, respectively) increases in absolute value while for $n = 4$ and 5 the wave function decreases, but have much larger absolute values in the barrier region

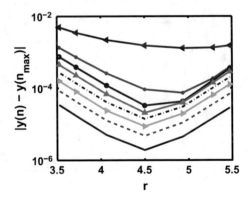

Fig. 8.15 The rate of convergence of the iterations for y described by Eq. (8.57), starting with the WKB approximation, for $k = 1.0855$, located in resonance #1. The maximum number of iterations is $n_{max} = 8$. The top curve (blue with triangles) represents the absolute value of the difference between $\tilde{y}_{WKB}$ and $\tilde{y}(n_{max})$, and the successively lower curves represent $|\tilde{y}(n) - \tilde{y}(n_{max})|$ for $n = 1, 2, \ldots, 7$. The decreasing differences of $y(n)$ with $y(n_{max})$ show that the iterations are converging

amplitude equal to 1 near $r = 100$, i.e., $\psi_C(r \to 100) = \sin(kr + \phi)$. The chosen values of r_1 and r_2 are 4.1 and 4.3, and the procedure is repeated for several values of k in the resonance region, given by

$$k(n_k) = 1.085267870 + (n_k - 1) \times 10^{-8}, \ n_k = 1, 2, \ldots, 10. \quad (8.62)$$

Fig. 8.16 The phase $\Phi(r)$ obtained from the amplitude y according to Eq. (8.49), for $k = 1.0855$, located in resonance #1. The difference between the WKB value and the result after 8 iterations ($n_{max} = 8$) is not visible in this figure

Fig. 8.17 The amplitude $\tilde{y}$ for $k = 1.0855$ in the barrier region. The result after 8 iterations clearly differs from the WKB approximation

Fig. 8.18 The wave functions $\psi^{(\pm)} = \tilde{y}\,\exp(\pm\Phi)$ as a function of the location r in the barrier region for $k = 1.0855$. These lines show the results obtained after 8 iterations

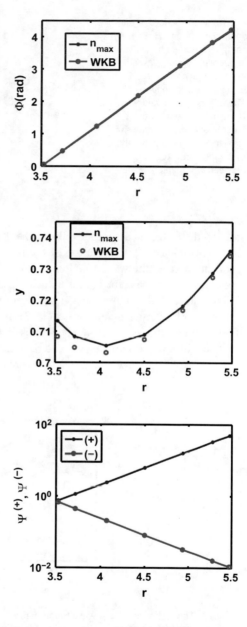

The resulting values of the Ph-A wave functions are illustrated by means of the discrete symbols in Figs. 8.12 and 8.13. The agreement with the S-IEM wave functions takes the form of the continuous lines in these figures and one can see that the calculation gives a very satisfactory result. However, even though the wave functions in the barrier region are very different from each other over the various $k-$ values, asymptotically the wave functions are again very similar. The corresponding phase

Fig. 8.19 The values of the coefficients A and B for the energies specified by the integers n_k, denoted as nk on the x-axis, according to Eq. (8.62), for resonance #1. The results for the values of B are shown by the almost horizontal line. The values of A change most dramatically in the resonance region between $n = 1$ and $n = 4$

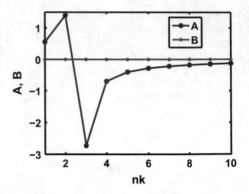

shifts ϕ differ from each other only in the 4th decimal place. This is not the case for the resonance #2 near the top of the barrier in Fig. 8.14, where for the various resonant energies the wave functions differ strongly and asymptotically from each other, nor is it the case for the resonance illustrated in Chap. 6.

The values of the coefficients A and B for resonance 1 are displayed in Fig. 8.19 for the set of energies given by Eq. (8.62). It is clear that the resonance occurs for energies between $n = 1$ and 4. The values of B are on the order of 10^{-5}, and are all positive with exception for $n = 1$ and $n = 2$, for which they are negative.

The error of the Ph-A wave function is displayed in Fig. 8.20.

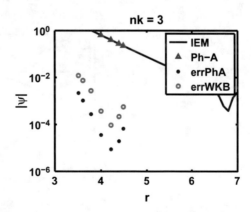

Fig. 8.20 The S-IEM and Ph-A wave function results are given by the solid line and the triangular symbols, respectively, for the wave number $k = 1.085267890$ in resonance #1. The potential is described by Eqs. (6.15) and (8.61). The errors of the WKB approximation and of the Ph-A result for the 8th iteration are given by the open and closed discrete symbols respectively, at the positions of the support points in the interval $3.5 \leq r \leq 4.5$ for an expansion of y in terms of seven Chebyshev polynomials

It can be seen from Fig. 8.20 that after 8 iterations, the error of the Ph-A wave function is smaller than that of the WKB wave function by approximately one order of magnitude.

8.4.4.2 A Non-resonant Case

In order to illustrate a case of a wider barrier, we have chosen the value $k = 0.1$. In this case the turning points are $T_1 \simeq 2$ and $T_2 \simeq 27$, but the region of convergence of the Ph-A iterations is contained approximately in the interval $4 \leq r \leq 8$. The number of Chebyshev functions in Eq. (8.11) is 11, and the convergence up to 8 iterations performed with Eq. (8.57) is illustrated in Fig. 8.21.

The dependence of the amplitude $\tilde{y}$ on r is illustrated in Fig. 8.22, and the corresponding phase $\Phi(r)$ is shown in Fig. 8.23.

The wave functions $\psi^{(\pm)}$ are illustrated in Fig. 8.24.

Since for this case in which $k = 0.1$, the local wave number is larger than that for the resonance case #1 for which $k \simeq 1.0853$, the amplitude $\tilde{y}$ is smaller (since the barrier is deeper), the phase Φ is larger, and the functions $\psi^{(\pm)}$ change more steeply with distance than for the $k \simeq 1.0853$ case, as can be easily understood from the WKB approximation. The absolute values of the Ph-A wave function points are displayed by means of the discrete triangles in Fig. 8.25, and the errors of these values are displayed by means of the circles.

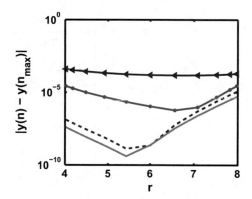

Fig. 8.21 The convergence of the iterations based on Eq. (8.57) for up to $n_{\max} = 8$ iterations is illustrated here for $k = 0.1$. The top line represents the absolute value of the difference between the WKB and result for $n_{\max} = 8$. The next lower line represents the difference between the $n = 1$ and the $n_{\max}$ result, and the next two lines represent the same for $n = 6$ and 7. The number of Chebyshev polynomials used in the expansion of $\tilde{y}$ is $N = 11$, and the radial interval is [4, 8]. Near $r = 8$ the iterations are barely converging, while between 4 and 7 they are converging adequately

Fig. 8.22 The amplitude y as a function of distance in the barrier region. The number of Chebyshev polynomials is 11 and $k = 0.1$. The results after the 8th iteration are visually indistinguishable from the WKB result. The symbols represent the Ph-A values at the support points and the lines are drawn so as to guide the eye

Fig. 8.23 The phase Φ as a function of distance in the barrier region. The number of Chebyshev polynomials is 11, and $k = 0.1$. The results after the 8th iteration are visually indistinguishable from the WKB result. The symbols represent the Ph-A values at the support points

Fig. 8.24 The two wave functions $\tilde{y}\,\exp(\pm\Phi)$ in the barrier region, after the 8th iteration, for the non-resonant value $k = 0.1$

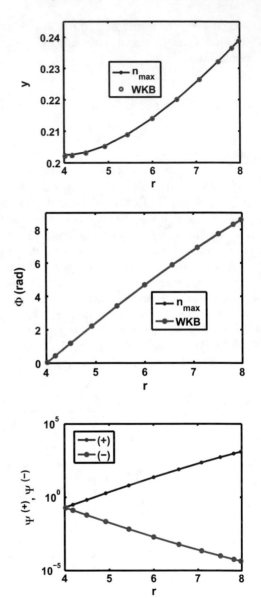

8.4.5   Numerical Examples for the 1/r Coulomb Case

Section 8.3.7 gives the conventional equation for the wave function $\chi(\rho)$ describing the point Coulomb case $\frac{d^2\chi}{d\rho^2} + \left[1 - \frac{2\eta}{\rho} - \frac{L(L+1)}{\rho^2}\right]\chi = 0$, and also gives the significance of the parameters $\bar{Z}$, η, ρ and of the phase shift θ_∞. In order to mitigate the

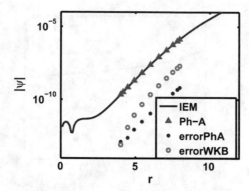

Fig. 8.25 The wave function in the case $k = 0.1$ for the Morse potential. The triangles illustrate the Ph-A results obtained with 11 Chebyshev polynomials in the barrier region $4 < r < 8$. The solid line represents the wave function obtained with the S-IEM method, and is normalized to unit amplitude near $r = 100$. The error points were obtained by comparing the values of the S-IEM wave functions with either the Ph-A values or the WKB values at the 11 Ph-A support points. The purpose of this graph is to show that the wave function increases exponentially in the barrier region, and that the WKB is not a bad approximation

singularities at the origin of r, the same "rounding" procedure described in Sect. 8.3.6 is introduced for the numerical examples. While in Sect. 8.3.7 the Coulomb potential is attractive, i.e., $\eta < 0$, the numerical examples below illustrate the applicability of the Phase-Amplitude method for the case in which the Coulomb potential is repulsive, i.e., $\eta > 0$.

8.4.5.1   Numerical Example #1

In this example the rounding parameter is $t = 2$, $\bar{Z} = 8$ ($\bar{Z}$ has units of inverse length), $V(r)$ (given by Eq. (8.60)) has units of inverse length squared), $k = 0.1$ (in units of inverse length), and $\eta = 40$. The convergence of the iterations of Eq. (8.54) or (8.55) requires that $\tilde{D} \ll V$. This is indeed the case for the present example, as illustrated by Fig. 8.26.

The convergence of successive values of $\tilde{y}_n$ is shown by a plot of $|\tilde{y}_n - \tilde{y}_{n_{max}}|$, $n = 1, 2, \ldots, n_{max-1}$ in Fig. 8.27. After $n = 8$ iterations, the resulting values of the amplitude and phase are illustrated in Figs. 8.28 and 8.29.

The resulting functions $\psi^{(+),(-)}$ are illustrated in Fig. 8.30.

As this figure shows, since the barrier is very long these wave functions change by many orders of magnitude, which for the case of $\psi^{(-)}$ would present substantial accuracy problems if calculated by methods other than the Phase-Amplitude.

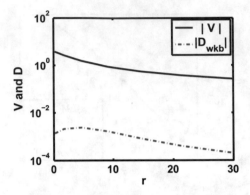

Fig. 8.26 The figure compares the potential V with the quantity $\tilde{D}$ for the rounded Coulomb potential described in this section. It shows that $|\tilde{D}|$ is several orders of magnitude smaller than $|V|$

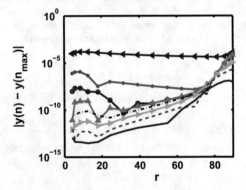

Fig. 8.27 The convergence of the iterations of the amplitude $\tilde{y}$ for the rounded (repulsive) Coulomb case with $\bar{Z} = 8$ and $k = 0.1$, with $4 \leq r \leq 90$, using 17 Chebyshev expansion polynomials. The corresponding value of η is 40, and the turning point is close to $r = 800$. The iterations are based on Eqs. (8.55) and (8.57). The top curve in blue with the triangle symbols pointing to the left represents the WKB approximation, and the subsequent curves in descending order correspond to the order of iteration index $n = 1, 2, \ldots, 7$, respectively

Fig. 8.28 The amplitude $\tilde{y}$ of the solution of Eq. (8.3) with the rounded repulsive Coulomb potential V given by Eq. (8.38), $\bar{Z} = 8$, $k = 0.1$, $N = 16$, and $4 \leq r \leq 90$

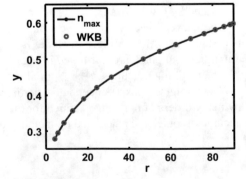

Fig. 8.29 The phase $\Phi(r)$ for the Coulomb parameters defined in Fig. 8.28. Since the phase increases by approximately $7 \times (2\pi)$, the corresponding wave function performs approximately 7 full oscillations

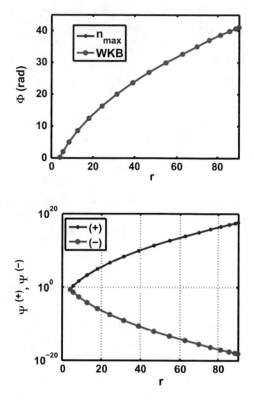

Fig. 8.30 The functions $\psi^{(+)}$ and $\psi^{(-)}$ as solutions of Eq. (8.3), obtained by means of the Ph-A method, with parameters described in Fig. 8.28. This graph shows the large difference between the $(+)$ and $(-)$ solutions, that were easily calculated by the Ph-A method, but that would be very difficult to calculate with a conventional wave function method

8.4.5.2 Numerical Example #2

The input parameters are the same as in Example #1, with the exception that now $\bar{Z} = 4$. If we were to make a comparison, the corresponding value of η is 20, which is the largest value for which a table of the regular Coulomb wave functions $F_0(\eta, \rho)$ values can be found in table 14.1 in Ref. [10]. In the Ph-A solution of Eq. (8.3) the range of r values is $0 \leq r \leq 60$ and the iterations are calculated by means of Eqs. (8.55) and (8.57). The resulting function $\psi^{(+)}(r) = \tilde{y}(r) \, e^{+\Phi(r)}$ is proportional to the regular (distorted) Coulomb function, but since it has not been continued beyond the outer turning point, the requirement that it must approach unit amplitude for $r \to \infty$ is not satisfied. Hence, $\psi^{(+)}(r)$ was normalized such that it agrees with $F_0(\eta, \rho)$ at $\rho = 3$, with the normalization factor $F_0(\eta, 3)/\psi^{(+)}(30) \simeq 8.72 \times 10^{-27}$. The comparison between the two functions is illustrated in Fig. 8.31. The good agreement shows that the Ph-A method gives reliable results for repulsive potentials that are as long ranged as the Coulomb potential.

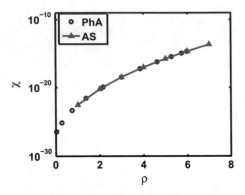

Fig. 8.31 Comparison of the $L = 0, \eta = 20$ regular Coulomb wave functions. The open circle results labeled with "PhA" are obtained with the Phase-Amplitude Method, and the results labeled with "AS" are taken from table 14.1 of Ref. [10]. The latter are connected by straight lines. The Ph-A results were normalized to the table 14.1 of Ref. [10] at the point $\rho = 3.0$

8.5   Conclusions

The Phase-Amplitude method for calculating a solution of the Schrödinger equation was shown to give reliable results for potentials as long ranged as the Coulomb potential. The method involves iterations required to handle the non-linearity of the equation for the amplitude y. In addition, the calculation is based on a spectral Chebyshev expansion of the amplitude. It is shown that the spatial applicability region of the method is restricted to the same region of applicability of the WKB approximation, meaning that the region has to be a certain distance from the turning points. The usefulness of the Ph-A description resides in the efficiency of the method, which is due to: (a) the small number of mesh points required in the calculation since neither the amplitude nor the phase change rapidly with distance; and (b) the numerical errors accumulate at a much smaller rate than when the full wave function is calculated directly, and hence leads to a stable accuracy out to large radial distances. Point (b) is illustrated in Chap. 7, by showing the large number of mesh-points needed on a finite difference Numerov calculation of the wave function, as compared to the few mesh-points in the Phase-Amplitude calculation presented here.

References

1. M.J. Seaton, G. Peach, Proc. Phys. Soc. **79**, 1296 (1962). https://doi.org/10.1088/0370-1328/79/6/127
2. W.E. Milne, Phys. Rev. **35**, 863 (1930)
3. F. Calogero, *Variable Phase Approach to Potential Scattering* (Academic, New York, 1967)
4. H. Jeffreys, B.S. Jeffreys, *Methods of Mathematical Physics* (Cambridge University Press, Cambridge, 1966)
5. H.A. Kramers, Z. Phys. **39**, 828 (1926)

6. D. Budker, D. Derek, F. Kimball, D.P. DeMille, *Atomic Physics: An Exploration through Problems and Solutions*, 2nd edn. (Oxford University Press, Oxford, 2008)
7. B. Wilson, C. Iglesias, M.H. Chen, J. Quant. Spectrosc. Radiat. Transf. **81**, 499 (2003)
8. D.J. Griffith, *Introduction to Quantum Mechanics*, 2nd edn. (Pearson-Prentice Hall, Upper Saddle River, 2005), p. 07458
9. G. Rawitscher, Comput. Phys. Commun. **191**, 33 (2015)
10. M. Abramowitz, I. Stegun (eds.), *Handbook of Mathematical Functions* (Dover, New York, 1972)
11. A.B. Ritchie, A.K. Bhatia, Phys. Rev. E **69**, 035402(R) (2004)
12. W.H. Press, S.A. Teukolsky, W.T. Vetterling, B.P. Flannery, Richardson extrapolation and the Bulirsch-Stoer method, *Numerical Recipes: The Art of Scientific Computing*, 3rd edn. (Cambridge University Press, New York, 2007). ISBN 978-0521880688
13. C.C. Clenshaw, A.R. Curtis, Numer. Math. **2**, 197 (1960)
14. R.A. Gonzales, J. Eisert, I. Koltracht, M. Neumann, G. Rawitscher, J. Comput. Phys. **134**, 134 (1997)
15. W.J. Olver, D.W. Lozier, R.F. Boisvert, C.W. Clark, *NIST Handbook of Mathematical Functions*, National Institute of Standards and Technology, U.S. Department of Commerce (Cambridge University Press, Cambridge, 2010)
16. G. Rawitscher, I. Koltracht, Comput. Sci. Eng. **7**, 58 (2005)
17. J.P. Boyd, *Chebyshev and Fourier Spectral Methods* (Dover Publications, Mineola, 2001)
18. A. Bar-Shalom, M. Klapisch, J. Oreg, Comput. Phys. Commun. **93**, 21 (1996)
19. G. Rawitscher, Phys. Rev. A **87**, 032708 (2013)
20. I. Simbotin, private communication
21. G. Rawitscher, Comput. Phys. Commun. **203**, 138 (2016)
22. Y. Cui et al., Phys. Rev. Lett. **119**, 203402 (2017)
23. G. Rawitscher, C. Merrow, M. Nguyen, I. Simbotin, Am. J. Phys. **70**, 935 (2002)
24. R.A. Gonzales, J. Eisert, I. Koltracht, M. Neumann, G. Rawitscher, J. Comput. Phys. **134**, 134 (1997)

Chapter 9
The Vibrating String

Abstract In this chapter we study the vibration spectra of both a homogeneous and an inhomogeneous string that is fixed at both ends. The description of the propagation of the waves on the inhomogeneous string requires the solution of a Sturm–Liouville eigenvalue equation. We obtain the solution by means of the Galerkin–Fourier expansion method and the spectral Green's function collocation method. The vibration frequencies are obtained as the eigenvalues of the corresponding matrices. We compare the advantages and disadvantages of both methods.

9.1 Summary and Motivation

A common procedure in physics or engineering consists in finding eigenvalues of a certain equation that describe the behavior of a physical system. These quantities are obtained as the eigenvalues of a matrix that describes the physical situation at hand. This is the method that will be described in the present chapter. For the Schrödinger equation the eigenvalues represent to bound state excitation energies for a given quantum system (atom or nucleus). In the case of a vibrating string the eigenvalues represent the modes of vibration of the string. The present chapter focuses on the vibrations of an inhomogeneous string. In this case one of the key equations (of the Sturm–Liouville type) that describes the waves propagating on the string is more complicated than the equation for a uniform string fixed at both ends, and hence offers a good computational test case. The corresponding eigenvalues can be obtained by various methods, one being based on conventional Fourier expansions for the solution of the differential equation, while another is based on the corresponding Lippmann-Schwinger integral equation. The purpose of the present chapter is to carry out both methods, and compare the merits and deficiencies of either one, while opening the way for the reader to examine other systems as well.[1]

---

[1] The material in this chapter is based mainly on Ref. [1].

© Springer Nature Switzerland AG 2018

G. Rawitscher et al., *An Introductory Guide to Computational Methods for the Solution of Physics Problems*,
https://doi.org/10.1007/978-3-319-42703-4_9

9.2   The Equations for the Inhomogeneous Vibrating String

Consider a stretched string of metal, clamped between two horizontal points P_1 and P_2 located in a horizontal plane. The distance between the fixed points is L, the mass per unit length ρ of the string is not a constant but varies along the string, as described below, and therefore the speed of propagation of the waves depends on the location x along the string. When a disturbance is excited along the string, the particles on the string vibrate in the vertical direction with a distribution of frequencies that are related to the eigenvalues of the S–L differential or integral equation.

Here $y(x, t)$ denotes the (small) displacement of a point on the string in the vertical direction away from the equilibrium position $y = 0$, for a given horizontal distance x of the point from the left end P_1, and at a time t. As can be shown, the wave equation is

$$\frac{\partial^2 y}{\partial x^2} - \frac{\rho}{T}\frac{\partial^2 y}{\partial t^2} = 0, \tag{9.1}$$

where T is the tension along the string. The derivation of this equation consists in mathematically dividing the string into small segments of length dL and mass $dm = \rho \times dL$, with a force acting on either end of the segment in approximately opposite directions. The two forces add vectorially and have a component pointing in the direction perpendicular to the tangent of the string element. The size of that component depends on the curvature $\partial^2 y/\partial x^2$ of the string element and accelerates the string element in the direction perpendicular to the tangent, according to the formulation of Newton's second law. After some additional approximations that depend on the size of the angle of the tangent with the horizontal direction, one obtains Eq. (9.1).

The method of separation of variables is used in order to find a solution to Eq. (9.1). We define a function $R(x)$ which is dimensionless, and which describes the variation of ρ with x according to

$$\rho(x) = \rho_0 \, R(x), \tag{9.2}$$

where ρ_0 is some fixed (or average) value of ρ. Defining a reference speed c according to

$$\frac{\rho_0}{T} = \frac{1}{c^2}, \tag{9.3}$$

the wave equation becomes

$$\frac{\partial^2 y}{\partial x^2} - \frac{1}{c^2}R(x)\frac{\partial^2 y}{\partial t^2} = 0. \tag{9.4}$$

By imposing a separation of variables, $y(x, t) = \psi(x)\, A(t)$, one obtains the two separate equations

$$\frac{d^2\psi(x)}{dx^2} + \Lambda R(x)\, \psi(x) = 0 \tag{9.5}$$

and

$$\frac{d^2A}{dt^2} = -\Lambda c^2 A. \tag{9.6}$$

Here Λ is an eigenvalue of Eq. (9.5) to be determined, which we assume to be positive, and ψ is an eigenfunction. A general solution for the corresponding time dependent Eq. (9.6) is $a\cos(wt) + b\sin(wt)$, with

$$w = c\sqrt{\Lambda}, \tag{9.7}$$

where the constants a and b are determined by the initial conditions.

The Eq. (9.5) is a Sturm–Liouville equation [2] (see also sections 3.9.3 and 5.5.2 in Ref. [3]) with an infinite set of eigenvalues Λ_n, $n = 1, 2, 3, \ldots$ and the corresponding eigenfunctions $\psi_n(x)$ form a complete set denoted as "Sturmians". The general solution of Eq. (9.4) can be expanded in terms of the Sturmian functions

$$y(x, t) = \sum_{n=1}^{\infty} [a_n \cos(\omega_n t) + b_n \sin(\omega_n t)] \, \psi_n(x), \tag{9.8}$$

where $w_n = c\sqrt{\Lambda_n}$. The objective is to calculate the functions $\psi_n(x)$ and the respective eigenvalues Λ_n as solutions of Eq. (9.5), with the boundary conditions that $y = 0$ for $x = 0$ and $x = L$,

$$\psi_n(0) = \psi_n(L) = 0, \; n = 1, 2, \ldots, \tag{9.9}$$

and that for $t = 0$

$$y(x, 0) = f(x) \text{ and } dy/dt|_{t=0} = g(x). \tag{9.10}$$

The constants a_n and b_n in Eq. (9.8) are given in terms of the initial conditions for the string, expressed by the functions $f(x)$ and $g(x)$. They are related to the initial displacement of the string from its equilibrium position $f(x)$ and the initial velocity $g(x)$ in terms of integrals of that displacement over the functions $\psi_n(x)$

$$a_n = \int_0^L f(x) \, \psi_n(x) \, dx; \quad b_n = \frac{1}{\omega_n} \int_0^L g(x) \, \psi_n(x) \, dx. \tag{9.11}$$

Exercises

9.1: Verify Eqs. (9.4)–(9.6).

9.2: Please derive Eq. (9.11).

9.3   The Homogeneous String

The case of the homogeneous string has been described in many textbooks and is reviewed here in order to provide a background for the case of the inhomogeneous string. For the homogeneous string the function $R(x) = 1$ becomes a constant, and the Sturmian expansion functions are given by the sine functions, i.e., $\psi_n(x) = \phi_n(x)$, with

$$\phi_n(x) = \sqrt{2/L}\,\sin(k_n x), \quad k_n = n(\pi/L), \quad n = 1, 2, 3, \ldots \qquad (9.12)$$

which vanish at both $x = 0$ and $x = L$. The corresponding eigenvalues become $\Lambda_n = k_n^2 = [n\pi/L]^2$. If one assumes that the initial displacement functions $f(x)$ and $g(x)$ of the string are given by the particular values

$$f(x) = x\sin[(\pi/L)x], \quad g(x) = 0, \qquad (9.13)$$

where $g(x) = 0$ implies that the initial velocity of all points on the string is zero, and if one assumes that

$$L = 1\,\text{m}, \quad c = 800\,\text{m/s}, \qquad (9.14)$$

then one can evaluate Eq. (9.11) for the coefficients a_n analytically (all the $b_n = 0$). One finds that all a_n vanish for n odd, with the exception for $n = 1$, for which

$$a_1 = -\frac{L^2}{4}\sqrt{\frac{2}{L}}. \qquad (9.15)$$

For n even, the corresponding result for a_n is

$$a_n = \frac{L^2}{\pi^2}\sqrt{\frac{2}{L}}\left[\frac{1}{(1+n)^2} - \frac{1}{(1-n)^2}\right], \quad n = 2, 4, \ldots. \qquad (9.16)$$

With the above results the sum (9.8), truncated at the upper values $n_{\max}$,

$$y^{(n_{\max})}(x, t) = \sum_{n=1}^{n_{\max}} [a_n\,\cos(\omega_n t) + b_n\,\sin(\omega_n t)]\,\phi_n(x), \qquad (9.17)$$

can be calculated. The result is displayed in Figs. (9.1) and (9.2).

Exercises

9.3: Please verify Eqs. (9.15)–(9.17).
9.4: Based on you results of Exercise 9-3, construct a code that will plot Figs. 9.1 and 9.2.

For $n \gg 1$, a_n will approach 0 like $(1/n)^3$, i.e., quite slowly. It is desirable to examine how many terms are needed in the numerical sum of Eq. (9.17) in order to get an accuracy of 4 significant figures in y. A good guess is that the sum of all terms

Fig. 9.1 Vibrations on the homogeneous string. The symbols in the plot mark the initial displacement of the string from its equilibrium position, given by Eq. (9.13). The numbers written next to each curve indicate the time, in units of L/c

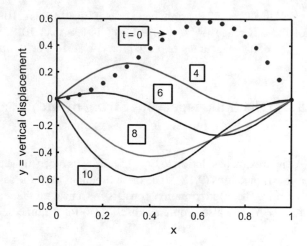

Fig. 9.2 Continuation from Fig. 9.1 of the time development of the vibrations of the string. One sees that the wave impulse was reflected from the left end of the string, and returns gradually to the initial form

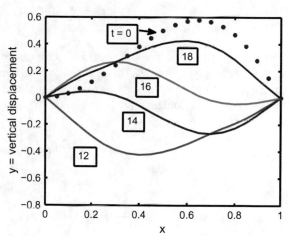

not included in the sum

$$\sum_{n_{max}+1}^{\infty} a_n \cos(\omega_n t) \simeq -4\frac{L^2}{\pi^2}\sqrt{\frac{2}{L}} \int_{n_{max}+1}^{\infty} \frac{1}{n^3} \cos\left(\frac{c\pi}{L}t\, n\right)\, dn \qquad (9.18)$$

should be less than $y_{max} \times 10^{-4}$. The integral in Eq. (9.18) is smaller than the integral $\int_{n_{max}+1}^{\infty}(1/n)^3 dn = (n_{max}+1)^{-2}/2$ (since the cos term produces cancellations), and one obtains the estimate

$$\left|\sum_{n_{max}+1}^{\infty} a_n \cos(\omega_n t)\right| < 2\frac{L^2}{\pi^2}\sqrt{\frac{2}{L}}(n_{max}+1)^{-2}. \qquad (9.19)$$

With $n_{max} = 50$ the right hand side of Eq. (9.19) is $\simeq 10^{-4}$. A numerical evaluation of the difference $|y^{(50)}(x, 0) - f(x)|$ is less than 10^{-5}, which confirms that with $n_{max} = 50$ the accuracy expected for $y^{(50)}(x, t)$ is better than $1 : 10^4$.

9.4 The Inhomogeneous String by Means of a Fourier Series

An approximate solution to Eq. (9.5) for ψ_n is proceed via the Galerkin method, using as basis functions (9.12) since these functions obey the same boundary conditions as the $\psi'_n s$. The approximation consists in truncating that expansion at an upper limit $\ell_{max} = N$, and also drop the subscript (n) for the time being

$$\psi^{(N)}(x) = \sum_{\ell'=1}^{N} d_{\ell'} \phi_{\ell'}(x). \tag{9.20}$$

By inserting expansion (9.20) into Eq. (9.5), remembering that $d^2\phi_\ell(x)/dx^2 = -k_\ell^2\phi_\ell(x)$, multiplying Eq. (9.5) by a particular function $\phi_\ell(x)$, integrating both sides of the equation over dx from $x = 0$ to $x = L$, and using the orthonormality of the functions $\phi_\ell(x)$, one obtains the matrix equation for the expansion coefficients d_ℓ, with $\ell = 1, 2, \ldots, N$

$$- k_\ell^2\, d_\ell + \Lambda \sum_{\ell'=1}^{N} R_{\ell,\ell'} d_{\ell'} = 0, \tag{9.21}$$

where

$$R_{\ell,\ell'} = \int_0^L \phi_\ell(x) R(x) \phi_{\ell'}(x)\, dx \tag{9.22}$$

are the matrix elements of the function R with respect to the basis functions ϕ_ℓ. This Eq. (9.21) can also be written in matrix form, where

$$\begin{pmatrix} k_1^2 & & & & \\ & k_2^2 & & & \\ & & k_3^2 & & \\ & & & \ddots & \\ & & & & k_N^2 \end{pmatrix} \begin{pmatrix} d_1 \\ d_2 \\ d_3 \\ \vdots \\ d_N \end{pmatrix} = \Lambda \begin{pmatrix} R_{1,1} & R_{1,2} & R_{1,3} & \ldots & R_{1,N} \\ R_{2,1} & R_{2,2} & R_{2,3} & \ldots & R_{2,N} \\ R_{3,1} & R_{3,2} & R_{3,3} & \ldots & R_{3,N} \\ \vdots & \vdots & \vdots & \ddots & \vdots \\ R_{N,1} & R_{N,2} & R_{N,3} & \ldots & R_{N,N} \end{pmatrix} \begin{pmatrix} d_1 \\ d_2 \\ d_3 \\ \vdots \\ d_N \end{pmatrix}, \tag{9.23}$$

or more succinctly

$$\hat{k}^2(d) = \Lambda\, R(d), \tag{9.24}$$

where a quantity in parenthesis indicates a $(N \times 1)$ column. Since all the k_ℓ's are positive, the matrix $\hat{k}^{-1}$ can be defined as

$$\hat{k}^{-1} = \begin{pmatrix} k_1^{-1} & & & \\ & k_2^{-1} & & \\ & & k_3^{-1} & \\ & & & \ddots \\ & & & & k_N^{-1} \end{pmatrix}. \tag{9.25}$$

Equation (9.24) is a generalized eigenvalue equation. It can be transformed into a simple eigenvalue Eq. (9.24) by defining

$$(u_n) = \hat{k}\,(d_n) \tag{9.26}$$

with the result

$$M_{Fourier}\,(u_n) = \frac{1}{\Lambda_s}(u_n); \quad n = 1, 2, \ldots, N, \tag{9.27}$$

where

$$M_{Fourier} = \hat{k}^{-1} R\,\hat{k}^{-1}. \tag{9.28}$$

The method described above is quite similar to the one presented in Ref. [3], section 6.7.5.

Exercise 9.5: Please confirm the validity of Eqs. (9.27) and (9.28), starting from Eq. (9.24).

The vectors (u_n) are the N eigenvectors of the $N \times N$ matrix $M_{Fourier}$, and $1/\Lambda_n$ are the eigenvalues. Furthermore, since R is a symmetric matrix, $M_{Fourier}$ is also symmetric. The eigenvectors of a symmetric matrix are orthogonal to each other, i.e., $(u_n)^T \cdot (u)_m = \delta_{n,m}$. Here T indicates transposition. However the vectors (d_n) are not orthogonal to each other, since $(d_n)^T \cdot (d_m) = (u_n)^T \cdot \hat{k}^{-2}(d_m)$.

In summary, the procedure is as follows (written in terms of MATLAB commands):

1. Choose an upper truncation limit N of the sum (9.20).
2. Calculate the matrix elements $R_{\ell,\ell'}$ so as to obtain the $N \times N$ matrix R.
3. Construct the matrix $M_{Fourier}$ from Eq. (9.28), and find the eigenvalues $(1/\Lambda_n)$ and eigenvectors (u_n), $n = 1, 2, \ldots, N$, by using the MATLAB eigenvalue command $[V, D] = eig(M)$. The output D is a diagonal matrix of the eigenvalues and V is a full matrix whose columns are the corresponding eigenvectors so that $M \times V = V \times D$. For example, the column vector $(u_n) = V(:, n)$.
4. If $(\Phi(x))$ is the column vector of the N basis functions $\phi_\ell(x)$, then $\psi(x)$ can be written as (the superscript (N) is dropped now)

$$\psi_n(x) = (u_n)^T \hat{k}^{-1} \cdot (\Phi(x)). \tag{9.29}$$

Please note that $(u_n)^T \hat{k}^{-1}$ is a row vector and $(\Phi(x))$ is a column vector, and hence $\psi_n(x)$ is a scalar.

5. In view of Eq. (9.29) the coefficients $a_n = (f|\psi_n)$ and $b_n = (g|\psi_n)$ can be written as

$$a_n = (u_n)^T \hat{k}^{-1} \cdot (f \, |(\Phi(x)))\,, \tag{9.30}$$

$$b_n = (u_n)^T \hat{k}^{-1} \cdot (g \, |(\Phi(x)))\,, \tag{9.31}$$

where $(f \, |(\Phi(x)))$ is the column vector of the integrals

$$(f | \, \phi_\ell) = \int_0^L f(x)\phi_\ell(x)dx, \; \ell = 1, 2, \ldots, N.$$

6. The final expression for $y(x, t)$ can be obtained by first obtaining the coefficients e_n

$$e_n(t) = (u_n)^T \times \hat{k}^{-1} \left[(f \, |(\Phi(x))) \cos(w_n t) + (g \, |(\Phi(x))) \frac{1}{w_n} \sin(w_n t) \right], \tag{9.32}$$

and then performing the sum

$$y(x, t) = \sum_{n=1}^N e_n(t)\psi_n(x) = (e)^T \times (\Psi). \tag{9.33}$$

In the above, (e) is the column vector of all e_n's, and likewise (Ψ) is the column vector of all ψ_n's. In the present discussion we limit ourselves to calculating the eigenvalues Λ_n.

9.4.1   A Numerical Example

Assuming that the mass per unit length changes quadratically with distance x from the left end of the string as

$$R(x) = 1 + F_0 \, x^2, \tag{9.34}$$

then the integrals (9.22) for the matrix elements $R_{\ell,\ell'}$ can be obtained analytically with the result

$$R_{\ell,\ell'} = 2 \cdot 2 \left(\frac{L}{\pi} \right)^2 (-1)^{\ell+\ell'} \left[\frac{1}{(\ell - \ell')^2} - \frac{1}{(\ell + \ell')^2} \right], \; \ell \neq \ell', \tag{9.35}$$

$$R_{\ell,\ell} = 1 + 2 L^2 \left[\frac{1}{3} - \frac{2}{(2\pi\ell)^2} \right], \; \ell = \ell'. \tag{9.36}$$

If c, $f(x)$ and $g(x)$ are the same as for the homogeneous string case, and if

$$L = 1\,\text{m}, \quad c = 800\,\text{m/s}, \quad f(x) = x\sin[(\pi/L)x], \quad g(x) = 0, \quad F_0 = 2, \qquad (9.37)$$

one obtains the results described below. The increase of R with x can be simply visualized with the choice (9.34). More realistic situations, such as the distribution of masses on a bridge, can be envisaged for future applications.

The numerical construction of the matrices R and $M_{Fourier}$ is accomplished in the MATLAB program *string_fourier.m* which in turn calls the function *inh_str_M .m*, using the input values

$$L = 1\,\text{m}, \quad c = 800\,\text{m/s}. \qquad (9.38)$$

The truncation value N of the sum Eq. (9.20) is set equal to either 30 or 60, and the corresponding dimension of the matrices $M_{Fourier}$ or R is $N \times N$. These values are chosen so as to examine the sensitivity of the eigenvalues to the size of the matrix $M_{Fourier}$.

The results for the eigenvalues Λ_n are shown in Fig. 9.3 and the corresponding frequencies are shown in Fig. 9.4. For comparison, the frequencies of the homogeneous string, i.e., for $R(x) = 1$, are shown by the open circles in Fig. 9.4. Since the inhomogeneous string is more dense at large values of x than the homogeneous one, the corresponding eigenfrequencies are correspondingly smaller because the vibrating pieces of the string have to carry a larger mass. It is noteworthy that the eigenfrequencies of the inhomogeneous string nearly fall on a straight line, which means that the frequencies are nearly equispaced, meaning that they nearly follow the same harmonic relationship as the ones in the homogeneous string. The physical explanation for this property has not been investigated here, but could be connected to the fact that the waves for the high indices have more nodes than for the low indices, and hence lead to better averaging in a variational procedure.

Fig. 9.3 The eigenvalues of the matrix $M_{Fourier}$, defined in Eq. (9.28). The quantity N indicates the truncation value of the sum in Eq. (9.20), that expands the string displacement eigenfunction $\psi_n(x)$ into the Fourier functions $\phi_\ell(x)$. The dimension of the matrix $M_{Fourier}$ is $N \times N$

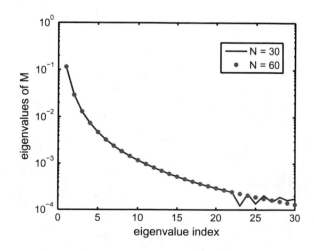

Fig. 9.4 The frequencies of
the vibration of the
inhomogeneous string given
in units of radians/s,
compared with the
frequencies of the
corresponding homogeneous
string. The higher
frequencies become
inaccurate when the
dimension of the matrix
$M_{Fourier}$ is too small

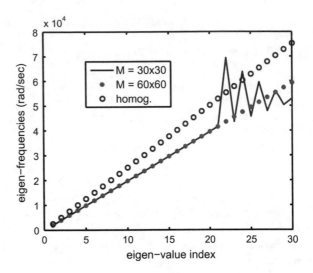

Nevertheless, near the fundamental frequency slight deviations from harmonicity
do occur, as illustrated in Fig. 9.5. Also, small deviations from harmonicity will also
be caused by other effects such as the stiffness of the string, and are not taken into
account here.

Figures 9.3 and 9.4 show that for the truncation value N of 30, the eigenvalues
become unreliable for $n \geq 22$. This is a general property of the high-n eigenvalues of
a matrix, and which can be overcome by using the iterative method that is described
in Chap. 10. Table 9.1 and Fig. 9.6 give a quantitative illustration of the dependence
of the eigenvalue on the truncation value N by comparing two eigenvalues for the
same n of the matrix $M_{Fourier}(30 \times 30)$ with those of $M_{Fourier}(60 \times 60)$.

Fig. 9.5 The deviation from
harmonicity as a function of
the eigenfrequency index for
two different
inhomogeneities. This
deviation is defined in terms
of the difference between
two neighboring frequencies
$d(n) = [w(n) - w(n - 1)]$
as
$\{d(n + 1)/d(n) - 1\} \times 100$.
The inhomogeneity is given
by $R(x) = 1 + F_0 x^2$ with F_0
either 2 or 4

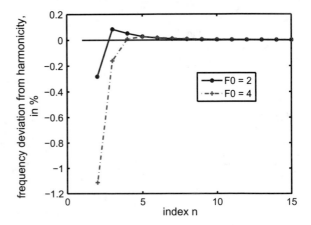

Table 9.1 Eigenvalues of the matrix $M_{Fourier}$ for two different dimensions $N \times N$

n	$N = 30$	$N = 60$
1	$1.614775590198150 \times 10^{-1}$	$1.6147755902115 \times 10^{-1}$
20	4.092×10^{-4}	$4.0933853097811 \times 10^{-4}$

Fig. 9.6 The dependence of the eigenvalues of the matrix $M_{Fourier}$ on the dimension $N \times N$ of the matrix. The y-axis shows the absolute value of the difference between two sets of eigenvalues, one for $N = 30$, the other for $N = 60$. Some numerical values are given in Table 9.1

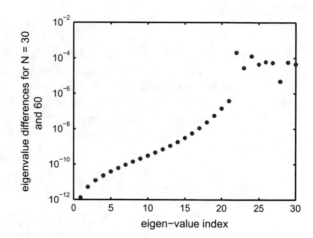

A more complete illustration of eigenvalues dependence on the dimension of the matrix is given in Fig. 9.6, again showing that for the high indices, the eigenvalues become unreliable.

9.5   The Integral Equation for the Inhomogeneous String

In the present section we introduce the following three major innovations: (a) according to Chap. 6, we transform the differential equation (9.5) into an integral equation [4, 5] in order to provide a different approach; (b) we replace the need to do overlap integrals by the Curtis Clenshaw [6] method given by Eq. (3.21), of obtaining the expansion coefficients; and (c) the basis functions are the Chebyshev polynomials for which the expansion series converges much faster than for the Fourier expansions.

The integral equation that is equivalent to the differential equation (9.5) is

$$\frac{1}{\Lambda}\psi(r) = -\int_0^L \mathscr{G}_0(r, r')\, R(r')\, \psi(r')dr'. \tag{9.39}$$

It is to be noted that this equation does not have a driving (or inhomogeneous) term, as is the case for Eq. (6.2) or (6.4), hence Eq. (9.39) is an eigenvalue integral equation. Furthermore, the Green's function $\mathscr{G}_0(r, r')$ is different. It is energy independent and is given by

$$\mathscr{G}_0(r, r') = -\frac{1}{L} F_0(r) G_0(r') \text{ for } r \leq r', \tag{9.40}$$

$$\mathscr{G}_0(r, r') = -\frac{1}{L} F_0(r') G_0(r) \text{ for } r \geq r', \tag{9.41}$$

where

$$F_0(r) = r; \quad G_0(r) = (L - r). \tag{9.42}$$

Both functions F_0 and G_0 obey the equation $d^2 F_0/dr^2 = 0$, $d^2 G_0/dr^2 = 0$ and they are linearly independent of each other. In addition, because $F_0(0) = 0$ and $G_0(L) = 0$, this Green's function incorporates the correct boundary conditions for the function ψ describing the vibrations of a string fixed at both ends.

The correctness of the boundary conditions can be verified as follows: Because of the separable nature of $\mathscr{G}_0$ the integral on the right hand side of Eq. (9.39) can be written as

$$\int_0^L \mathscr{G}_0(r, r') R(r') \psi(r') dr' = -\frac{1}{L} G_0(r) \int_0^r F_0(r') R(r') \psi(r') dr'$$
$$- \frac{1}{L} F_0(r) \int_r^L G_0(r') R(r') \psi(r') dr'. \tag{9.43}$$

In view of the fact that F_0 vanishes at $r = 0$ and G_0 vanishes at $r = L$, and hence $\int_0^L \mathscr{G}_0(r, r') R(r') \psi(r') dr'$ vanishes for both $r = 0$ and $r = L$, the functions ψ satisfy the boundary conditions. A proof that $\psi(r)$ defined by Eq. (9.39) satisfies Eq. (9.5) can be obtained by carrying out the second derivative in r of Eq. (9.43), as already suggested in Chap. 6.

The numerical solution of Eq. (9.39) is accomplished by first changing the variable r contained in the interval $[0, L]$ into the variable x contained in the interval $[-1, +1]$, which results in the transformed functions $\bar{\psi}(x)$, $\bar{\mathscr{G}}(x, x')$, and $\bar{R}(x')$. By expanding the unknown solution $\bar{\psi}(x)$ into Chebyshev polynomials

$$\bar{\psi}(x) = \sum_{n=1}^{N+1} a_n T_{n-1}(x), \tag{9.44}$$

Eq. (9.39) leads to a matrix equation in the coefficients a_n, as will now be shown. The method is very similar to the one described in Sect. 3 of Chap. 6, including the relevant matrices C, C^{-1}, S_L, and S_R. The support points ξ_i are the zeros of T_{N+1}, and the column vector of the coefficients a is denoted as $(a) = (a_1, a_2, \ldots, a_{N+1})^T$.

After a little algebra, and using the MATLAB commands notation, the final equation is

$$M_{IEM} = \frac{1}{2} \times C^{-1} \times M_3 \times DR \times C. \tag{9.45}$$

The factor $1/2$ comes from the transformation of coordinates from r to x, and where the term L was cancelled by the $(1/L)$ in Eq. (9.43), $DR = diag(R)$ is the diagonal matrix that contains the values of $R(\xi_i)$ along the main diagonal, where R is the inhomogeneity function defined in Eq. (9.2). Finally, M_3 is given by

$$M_3 = DG \times C \times S_L \times C^{-1} \times DF + DF \times C \times S_R \times C^{-1} \times DG. \qquad (9.46)$$

The first (second) term in Eq. (9.46) represents the first (second) term in Eq. (9.43), $DF = diag(F)$ and $DG = diag(G)$ represent the diagonal matrices having the values of $F(\xi_i)$ and $G(\xi_i)$ along the main diagonal, and ξ_i corresponds to the $N + 1$ Chebyshev support points described in Chap. 5.

A thumbnail explanation for Eq. (9.46) is as follows: the matrix M_{IEM} in Eq. (9.45) is applied to the column vector (a), the C in (9.45) transforms the (a) into the column vector (ψ), the factor DR together with the factor DG in (9.46) transforms (ψ) into $(G) \otimes (R) \otimes (\psi)$ (the symbol $\otimes$ means that in $(G) \otimes (R)$ each element of the vector (G) is multiplied by the corresponding element of the vector (R), and a new vector of the same length is produced), the additional factor C^{-1} produces the expansion coefficients of $(G) \otimes (R) \otimes (\psi)$, and the matrix S_L or S_R transforms these expansion coefficients to the expansion coefficients of the respective indefinite integrals, and so on.

9.5.1 A Numerical Example

For the purpose of this section, the upper truncation value $N + 1$ of the Chebyshev expansion (9.44) is denoted here as $N_{IEM} + 1$. After choosing a specific value for the number $N_{IEM} + 1$ a numerical value of the $(N_{IEM} + 1) \times (N_{IEM} + 1)$ matrix (9.45) is obtained, from which the eigenvalues $(1/\Lambda_n)$, $n = 1, 2, \ldots, N_{IEM} + 1$, can be calculated. The MATLAB computing times for the Fourier method for $N_{Fourier} = 30$ and 60 combined using the analytic expressions for the integrals needed to obtain the elements of the matrix R is $0.91\,\mathrm{s}$, while the computing time for the S-IEM matrix method for all three $N_{IEM} = 30, 60$, and 90 values combined is $0.75\,\mathrm{s}$. Hence the S-IEM method is comparable in complexity to the Fourier expansion method, provided that the Fourier overlap integrals (9.22) are known analytically. However, a disadvantage of the S-IEM for the present application is that some eigenvalues are spurious. Their occurrence can be recognized because they change with the value of N_{IEM} and do not correspond to the eigenvalues of $M_{Fourier}$.

The accuracy of these two matrix methods is illustrated in Fig. 9.7, based on the iterative method used as an accuracy benchmark described in Chap. 8. As explained in there, this method gives an accuracy of $1 : 10^{11}$ for the eigenvalues regardless of the value of the eigenvalue index n, given that the result is not based on the (possibly unreliable) eigenvalues of a matrix.

Figure 9.7 shows that the accuracy of the S-IEM matrix method is considerably higher than the Fourier matrix method for the low values of n, but it is not as monotonic

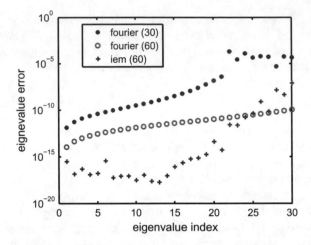

Fig. 9.7 Accuracy of the eigenvalues of $M_{Fourier}$ in Eq. (9.27) and M_{IEM} in Eq. (9.45), for various values of their dimension $(N + 1) \times (N + 1)$. The value of N is indicated in parenthesis on the legend. For the Fourier method, N is the number of basis functions ϕ_ℓ used to expand the Sturm–Liouville eigenfunctions, and for the S-IEM method, $N + 1$ is the number of Chebyshev polynomials used in the expansion, which is also equal to the number of support points in the interval $[0, L]$. The accuracy of the matrix eigenvalues is obtained by comparison with a highly accurate result of 1 part in 10^{11} obtained by an iterative method to be described in Chap. 10

as the latter. The figure also shows that the accuracy of both matrix methods depends sensitively on the dimension N of their respective matrices M. If the Fourier integrals could not be done analytically, then a numerical evaluation of these integrals would be less precise and more time intensive. Therefore, the use of the S-IEM Green's function method would be a better option.

9.6   Summary and Conclusions

The main aim of this chapter is to introduce the spectral Green's function Colloca-tion method as an alternative to obtain eigenvalues and eigenfunctions of a Sturm–Liouville equation. This method is compared with a Fourier Galerkin expansion method, for the purpose of clarifying the implementation of both methods. The example that we use here for the application of these methods is based on the anal-ysis of the vibration of an inhomogeneous string in the formalism of separation of variables. Applications of these methods to other problems, such as the solution of the Schrödinger equation, to the heat propagation equation, or to diffusion equa-tions in biology are of course quite possible in spite of the present focus on the inhomogeneous string equation.

Apart from the methodological difference between the Galerkin and the Colloca-tion approaches, both methods have in common the characteristic that the functions obtained are eigenfunctions of a matrix, whose dimension depends on the size of each of the respective expansion basis. Since the sizes of the spectral Collocation method are expected to be smaller (because spectral expansions converge rather rapidly) it is argued that the spectral Green's function method may be preferable for most applications, even though this method is more complex than the Galerkin–Fourier one.

References

1. G. Rawitscher, J. Liss, The vibrating inhomogeneous string. Am. J. Phys. **79**, 417–427 (2011)
2. M.L. Boas, *Mathematical Methods in the Physical Sciences*, 2nd edn. (Wiley, New York, 1983), Problem 24 on p. 540; D.A. McQuarrie, *Mathematical Methods for Scientists and Engineers* (University Science Books, Sausalito, 2003), p. 687, ff
3. B.D. Shizgal, *Spectral Methods in Chemistry and Physics. Applications to Kinetic Theory and Quantum Mechanics* (Springer, Dordrecht, 2015)
4. R.A. Gonzales, J. Eisert, I. Koltracht, M. Neumann, G. Rawitscher, J. Comput. Phys. **134**, 134–149 (1997); R.A. Gonzales, S.-Y. Kang, I. Koltracht, G. Rawitscher, J. Comput. Phys. **153**, 160–202 (1999)
5. G. Rawitscher, I. Koltracht, Description of an efficient numerical spectral method for solving the Schrödinger equation. Comput. Sci. Eng. **7**, 58–66 (2005); G. Rawitscher, Applications of a numerical spectral expansion method to problems in physics; a retrospective, *Operator Theory, Advances and Applications*, vol. 203 (Birkäuser Verlag, Basel, 2009), pp. 409–426
6. C.C. Clenshaw, A.R. Curtis, Numer. Math. **2**, 197 (1960)

Chapter 10
Iteratively Calculated Eigenvalues

Abstract In this chapter, we study an accurate iterative method used to calculate eigenvalues of a second order differential equation. The method was introduced by Hartree for calculating energy eigenvalues of the Schrödinger equation for atomic systems. We show that the use of the spectral S-IEM method to obtain the wave functions required in the procedure can improve the precision to $1 : 10^{-11}$. The numerical application that we present here is again based on the vibration frequencies of an inhomogeneous string as presented in Chap. 9. To close the chapter, we propose a project which applies the method to the case of an exponential attractive potential.

10.1 Summary and Motivation

The iterative method which we are going to work on in this chapter was first introduced by Hartree [1] in the 1950s as a solution for the calculation of energy eigenvalues of the Schrödinger equation for atomic systems. The method is not based on the eigenvalues of a numerical matrix, given that it becomes unreliable for a sufficiently high eigenvalue index as shown in Chap. 9, but is based on an iteration scheme that is applicable to second order differential equations no matter how high the value of the index may be. The drawback is that if the iterations are to converge, the starting value of the eigenvalue search has to be sufficiently close to the final result, and has to lie within the valley of convergence of the iteration. In the present formulation, we improve the precision of Hartrees method through the use of the spectral expansion method [2, 3] (S-IEM). It was applied to the energy eigenvalue of the very tenuously bound Helium-Helium dimer [4], and was found to be very reliable. The basic iteration method will be described for the example of a vibrating inhomogeneous string, whose equation, $d^2\psi/dr^2 + \Lambda R(x)\,\psi(x) = 0$ (9.5), is given in Chap. 9. The procedure is also applicable to other situations, such as finding the Sturmian functions associated with a Schrödinger equation with an arbitrary potential function. In that case the equation is $(-d^2\psi/dr^2 - k^2\psi) + \Lambda V(r)\psi(r) = 0$, the energy k^2 is a known input number, $V(r)$ is the potential function, and Λ is the eigenvalue to be calculated. The Sturmian case will be treated in Chap. 11 [5, 6]. Here the emphasis

© Springer Nature Switzerland AG 2018
G. Rawitscher et al., *An Introductory Guide to Computational Methods for the Solution of Physics Problems*,
https://doi.org/10.1007/978-3-319-42703-4_10

is to obtain the value of Λ iteratively, and not as the eigenvalue of a matrix, as for instance is presented in Ref. [7].

The basic requirement in all these cases is that the pre-assigned boundary conditions of the eigenfunctions are satisfied, which becomes possible only if the eigenvalue has the correct value. The iterative procedure starts with a guessed eigenvalue, and the corresponding wave function is propagated from the outside (large r) inwards, and from the origin (small r) outwards towards a certain matching point r_I in between. From the mismatch at this point of the two wave functions, together with the second order derivative equation that these functions obey, an improved guess for the eigenvalue is calculated. The iterations are stopped once a certain accuracy is reached.

The material in this chapter is based mainly on Ref. [4].

10.2   The Method for a Vibrating String

The version described below focuses on finding the eigenfunctions ψ and eigenvalues Λ of the equation

$$\frac{d^2\psi(r)}{dr^2} + \Lambda R(r)\,\psi(r) = 0. \tag{10.1}$$

Here R is a given function of distance r, which was introduced in Chap. 9 to express the dependence of the density of the material of a vibrating string as a function along the distance on the string. When appropriately modified, the method is also suitable for finding the eigenfunctions in more general Sturm–Liouville equations. The iterative method for solving for Λ in Eq. (10.1) is to obtain a series of functions ψ_n and eigenvalues Λ_n, $n = 1, 2, \ldots$, that converge to ψ and to Λ.

The method is as follows [1]. For a slightly wrong value Λ_1 of Λ there is a slightly wrong function ψ_1 that obeys the equations

$$\frac{d^2\psi_1(r)}{dr^2} + \Lambda_1 R(r)\,\psi_1(r) = 0. \tag{10.2}$$

This function does not satisfy the boundary conditions at both $r = 0$ and $r = L$ unless it has a discontinuity at some point r_I, contained in the interval $[0, L]$. To the left of r_I the function ψ_1 that vanishes at $r = 0$ and obeys Eq. (10.2) is called $Y_1(r)$, and to the right of r_I it is called $\mathfrak{k} * Z_1(r)$, and vanishes at $r = L$. Here $\mathfrak{k}$ is a normalization factor chosen such that $Y_1(r_I) = \mathfrak{k} \, Z_1(r_I)$. Both these functions rigorously obey Eq. (10.2) in their respective intervals and are obtained by solving the domain-limited integral equations

$$Y_1(r) = F_0(r) - \Lambda_1 \int_0^{r_I} \mathcal{G}_0(r, r')R(r')Y_1(r')dr', \quad 0 \le r \le r_I, \tag{10.3}$$

and

$$Z_1(r) = G_0(r) - \Lambda_1 \int_{r_l}^{L} \mathcal{G}_0(r, r') R(r') Z_1(r') dr', \quad r_l \leq r \leq L. \tag{10.4}$$

The functions F_0 and G_0 are independent of Λ and are defined by Eq.(9.42) in Chap.9, while the Green's function $\mathcal{G}_0(r, r')$ is defined in Eqs.(9.40) and (9.41). These integral equations are not eigenvalue equations since they contain the driving terms F_0 and G_0. However, because the second derivatives of these functions are zero, their presence does not prevent Y_1 and Z_1 from obeying Eq.(10.2) in their respective domains. Furthermore, since the variation of R in the radial domain $[0, L]$ is small, Eqs.(10.3) and (10.4), can be solved with the spectral S-IEM method, meaning that the intervals $[0, r_i]$ and $[r_i, L]$ need not be subdivided further.

The iteration from Λ_1 to a value closer to the true Λ proceeds as follows. The function Y_1 obeys

$$\frac{d^2 Y_1(r)}{dr^2} + \Lambda_1 R(r) Y_1(r) = 0, \ 0 \leq r \leq r_l. \tag{10.5}$$

We multiply Eq.(10.5) with $\psi(r)$ and multiply Eq.(10.1) with $Y_1(r)$, subtract one from the other, and integrate from $r = 0$ to $r = r_l$. As a result we find that $\int_0^{r_l} (Y_1''\psi - \psi''Y_1) dr' = (Y_1'\psi - \psi'Y_1)_{r_l} = (\Lambda - \Lambda_1) \int_0^{r_l} Y_1\psi dr'$. Here a prime denotes the derivative with respect to r. A similar procedure applied to Z_1 in the interval $[r_l, L]$ yields $-\ell (Z_1'\psi - \psi'Z_1)_{r_l} = (\Lambda - \Lambda_1) \int_0^{r_l} \ell Z_1\psi dr'$. Remembering that $\ell Z_1 = Y_1$ for $r = r_l$, and dividing each of these results by $(Y_1\psi)_{r_i}$ and $(Z_1\psi/\ell)_{r_i}$, respectively, and then adding them, one obtains

$$\Lambda - \Lambda_1 = \frac{(Y'/Y - Z'/Z)_{r_l}}{\frac{1}{(Y_1\psi)_{r_l}} \int_0^{r_l} Y_1 R\psi dr' + \frac{1}{(Z_1\psi)_{r_l}} \int_{r_i}^{L} Z_1 R\psi dr'}. \tag{10.6}$$

This result is still exact, but the exact function ψ corresponding to the exact value of Λ is not known. The iterative approximation occurs by replacing ψ in the first integral in the denominator by Y_1, and by ℓZ_1 in the second integral, and by replacing $\psi(r_l)$ in the denominators of each integral by either $Y_1(r_l)$ or by $\ell Z_1(r_l)$. The normalization factor ℓ cancels and the final result is

$$\Lambda_2 = \Lambda_1 + \frac{(Y'/Y - Z'/Z)_{r_l}}{\frac{1}{Y_1^2(r_l)} \int_0^{r_l} Y_1^2 R dr' + \frac{1}{Z_1^2(r_l)} \int_{r_i}^{L} Z_1^2 R dr'}. \tag{10.7}$$

In the above, Λ was replaced by Λ_2 as a better approximation to Λ than Λ_1. The iteration proceeds by replacing Λ_1 in the above equations by the new value Λ_2, and so on.

The derivatives in the numerator of Eq.(10.7) can be obtained without loss of accuracy by making use of the derivatives of Eqs.(10.3) and (10.4)

$$Y_1'(r) = F'(r) + \frac{\Lambda_1}{L}G'(r) \int_0^r F(r')R(r')\, Y_1(r')dr' +$$

$$\frac{\Lambda_1}{L}F'(r) \int_r^{r_I} G(r')R(r')\, Y_1(r')dr', \tag{10.8}$$

and

$$Z_1'(r) = G'(r) + \frac{\Lambda_1}{L}G'(r) \int_{r_I}^r F(r')R(r')\, Z_1(r')dr' +$$

$$\frac{\Lambda_1}{L}F'(r) \int_r^L G(r')R(r')\, Z_1(r')dr', \tag{10.9}$$

with the result at $r = r_I$

$$Y_1'(r_I) = 1 - \frac{\Lambda_1}{L} \int_0^{r_I} r'R(r')\, Y_1(r')dr', \tag{10.10}$$

and

$$Z_1'(r_I) = -1 + \frac{\Lambda_1}{L} \int_{r_I}^L (L - r')R(r')\, Z_1(r')dr'. \tag{10.11}$$

In the present formulation, the dimensions of Λ are ℓ^{-2} and the dimension of F_0, G_0, Y and Z are ℓ, where ℓ represents a unit of length, and R has no dimension. As noted above, the derivatives with respect to r of the functions Y or Z or ψ are not obtained as the difference between two adjoining positions, but rather in terms of the analytically known derivatives of F and G, together with integrals over Y or Z or ψ according to Eqs. (10.8) and (10.9). In the S-IEM formulation these integrals can be obtained with the same spectral precision as the calculation of the functions Y or Z or ψ [3], hence there is no loss of accuracy either for the evaluation of Eq. (10.7), or for the calculation of Λ, which can be set to $1 : 10^{11}$. However, it is important to start the iteration with a guessed value of Λ that lies within the valley of convergence of Eq. (10.7). These initial values can be obtained from the eigenvalues of the matrix $M_{Fourier}$ described in Chap. 9, or from a method described in Ref. [4] amongst others.

10.2.1   Numerical Example for the Iterative Method

For the results obtained in the present section the value of $L = 1m$ and the function R is given by

$$R(r) = 1 + 2r^2, \ 0 \le r \le L.$$

Some of the values for Λ_n obtained to an accuracy of $1 : 10^{11}$ by means of the iterative method given by Eq. (10.7) are listed in Table 10.1, so as to serve as benchmark results

Table 10.1 Eigenvalues of Eq. (10.1) obtained iteratively with Eq. (10.7)

n	Λ_n	n	Λ_n
1	$1.61477559021 \times 10^{-1}$	26	$2.42220326385 \times 10^{-4}$
2	$4.06257259855 \times 10^{-2}$	27	$2.24611142229 \times 10^{-4}$
3	$1.81281029690 \times 10^{-2}$	28	$2.08854647313 \times 10^{-4}$
4	$1.02131986136 \times 10^{-2}$	29	$1.94699775697 \times 10^{-4}$
5	$6.54130338213 \times 10^{-3}$	30	$1.81936592475 \times 10^{-4}$

for comparisons with future methods. The starting values Λ_1 for each n are the results of the Fourier method described in Chap. 9 with $N_{Fourier} = 60$. The iterations were stopped when the change $\Lambda_2 - \Lambda_1$ became less than 10^{-12} (usually three iterations were required), and $tol = 10^{-11}$ for the solution of Eqs. (10.3) and (10.4).

The error of the functions Y and Z is given according to Eq. (4.25) in Chap. 4 by the size of the high order Chebyshev expansion parameters. For the tol parameter of 10^{-11} their values stay below 10^{-11}, as is shown in Fig. 10.1. Since there is no loss of accuracy in evaluating the various terms in Eq. (10.7), the error in the iterated eigenvalues Λ is also given by Fig. 10.1. In order to achieve this type of error for each eigenvalue Λ_n, $n = 1, 2, \ldots, 30$, the number N of Chebyshev polynomials used for the spectral expansion of the functions Y and Z for the solution of their respective integral equations was increased adaptively by the computer program. It was found that for $n = 1, N = 16$; for $n = 2$ to $6, N = 24$; for $n = 7$ to $23, N = 24$; and for $n = 18$ to $30, N = 54$. This procedure of increasing N is different from the

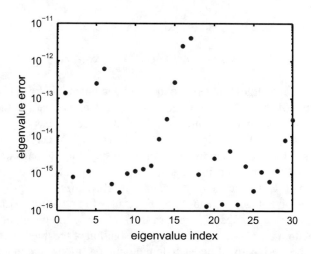

Fig. 10.1 The y-axis shows the absolute value of the mean square average of the three last Chebyshev coefficients in the expansions of the functions Y and Z. As discussed in the text, the error of the eigenvalues Λ is also given by the y-axis. The number N of expansion Chebyshev polynomials was increased adaptively as the eigenvalue index n increased. The "jumps" in the values of these errors is due to the transition from one value of N to a suddenly larger value, as is explained in the text

procedure used in Ref. [4], where N was kept constant and the number of partitions was increased adaptively. The use of the latter method was required because of the long range (3000 units of length) of the He–He wave functions.

10.3   Calculation of an Energy Eigenvalue

The equation to be solved can be written in the dimensionless form

$$-\frac{d^2\psi}{dr^2} + (V + \kappa^2)\psi = 0 \tag{10.12}$$

where V is the potential and the negative energy is given by $-\kappa^2$. Here κ is a positive number whose value is to be found iteratively. For atomic physics applications $r = \bar{r}/a_0$ is the relative distance in units of Bohr, and V and κ^2 are given in atomic energy units. The L–S equation that is the equivalent to Eq. (10.12) is

$$\psi(r) = \int_0^{r_{max}} \mathscr{G}(\kappa, r, r')V(r')\psi(r')dr', \tag{10.13}$$

where

$$\mathscr{G}(\kappa, r, r') = -\frac{1}{\kappa}F_\kappa(r_<)G_\kappa(r'_>), \tag{10.14}$$

with $r_<$ and $r'_>$ being the least and the largest values of r and r', and

$$F_\kappa(r) = \sinh(\kappa r), \quad G_\kappa(r) = \exp(-\kappa r), \tag{10.15}$$

and where r_{max} is the largest value beyond which the potential V can be ignored.

Equation (10.13) is an eigenvalue equation that does not satisfy the boundary condition that $\psi(r)$ decay exponentially at large distances unless the wave number κ has the correct value. The method of finding the correct value of κ is to start with an initial guess κ_s for κ, and then to divide the corresponding (wrong) wave function into an "out" and an "in" part, in order to match the two at an intermediary point r_I. The "out" part ψ_O is obtained by integrating (10.13) from the origin out to an intermediate radial distance r_I, and ψ_I is the result of integrating (10.13) from the upper limit of the radial range r_{max} inward to r_I. Because the potential V can vary substantially within the two radial domains, the solution of the integral equations for ψ_O and ψ_I is obtained by dividing each domain into partitions, and then proceeding by the finite element spectral integral equation method (S-IEM), described in Refs. [2, 3]. The function ψ_O is renormalized so as to be equal to ψ_I at $r = r_I$ and its value at $r = r_I$ is denoted as ψ_M. The derivatives with respect to r at $r = r_I$ are calculated as described below, and are denoted as ψ'_O and ψ'_I respectively. By a method similar to the one which led to Eq. (10.7), the new value of the wave number κ_{s+1} is given in terms of these quantities as

$$\kappa_{s+1} = \kappa_s - (Iter)_s, \tag{10.16}$$

where

$$(Iter)_s = \frac{1}{2\kappa_s} \frac{\psi_M (\psi_0' - \psi_I')_M}{\int_0^{T_M} \psi_0^2 dr + \int_{T_M}^T \psi_I^2 dr}. \tag{10.17}$$

The factor $2\kappa_s$ arises from the approximation of $\kappa^2 - \kappa_s^2$ by $(\kappa - \kappa_s)(\kappa + \kappa_s) \simeq (\kappa - \kappa_s)(2\kappa_s)$, and the function ψ is approximated by ψ_0 or ψ_I in the respective integration domains. The calculation of the functions ψ_0 and ψ_I for the case of negative energies involves a renormalization at each partition thus avoiding that the two functions F_κ and G_κ become too disparate (non-equal magnitudes) from each other so as not to lose accuracy. These renormalizations, and the explanation of how to propagate the wave function from one partition to the subsequent one, are detailed in Ref. [4].

The search for an appropriate starting value of κ_s is based on the boundary condition requirement that the partitions J close to the end point r_{max} do not contain a dominant exponentially increasing component. In each partition the wave function $\psi^{(J)}(r)$ is composed of the linear combination of the two functions $Y_{\kappa_s}^{(J)}$ and $Z_{\kappa_s}^{(J)}$, each of which obeys the negative energy equivalent of Eq. (6.2), driven either by F_{κ_s} or G_{κ_s}

$$\psi^{(J)}(r) = A_{\kappa_s}^{(J)} Y_{\kappa_s}^{(J)}(r) + B_{\kappa_s}^{(J)} Z_{\kappa_s}^{(J)}(r). \tag{10.18}$$

The boundary condition requirement above is satisfied if the coefficient $A_{\kappa_s}^{(J)}$ is going through a zero as κ_s increases. Hence a grid of κ_s values is set up, and the particular values of κ_s for which $A_{\kappa_s}^{(J)}$ goes through zero, for a fixed value J of the partition are marked as the initial values for each energy level.

For the calculation of the binding energy of the He–He diatom [4] the value of r_{max} is 3,000, at which point the potential has a value of $\simeq 6 \times 10^{-9}$, and the tolerance parameter is $tol = 10^{-12}$, hence the accuracy of the final wave number eigenvalue κ is expected to be better than 10^{-10}. The rate of convergence of the iterations is shown in Table 10.2. Comparison with results contained in the literature is found in Ref. [4].

Table 10.2 Convergence of the iterations for the wave number	s	$\kappa_s \ (a_0)^{-1}$	$Iter_s$ (from (10.17))
	0	3.0×10^{-3}	$-2.5002592843 \times 10^{-3}$
	1	$5.5002592823 \times 10^{-3}$	$-1.0967998971 \times 10^{-5}$
	2	$5.5112272813 \times 10^{-3}$	$-2.0105203008 \times 10^{-10}$
	3	$5.5112274823 \times 10^{-3}$	$-4.9700035857 \times 10^{-16}$

10.3.1 Project 10.1 (Difficult)

Reconsider the attractive exponential potential defined in Chap. 6,

$$V(r) = V_0 \times \exp(-r); \quad V_0 = -5. \tag{10.19}$$

(a) Examine whether one or more bound states exist in this potential. If they do not exist, increase the value of $|V_0|$ until a bound state is found. Hint: Set up a Green's function matrix, as was done in Chap. 9 for the vibrating string, and examine the eigenvalues for various values of the negative energy (i.e., assign negative values to k^2).

(b) Using analytical methods, try to obtain values of bound state energies.

(c) Using the iterative method described in Chap. 10, refine the accuracy of the energy eigenvalues found in part (a).

10.4 Summary and Conclusions

The iterative method of obtaining eigenvalues and eigenfunctions of a second order linear differential equation, originally given by Hartree [1], is implemented in this Chapter by combining the method with a spectral expansion for the calculation of the required wave functions. The advantages of this method include that the iterations converge very quickly to high accuracy once a good initial value for the start of the iterations is found, and that the final values do not depend on the eigenfunctions of a particular matrix, since the latter become unreliable for high values of the eigenvalue index.

References

1. D.R. Hartree, *Numerical Analysis* (Clarendon Press, Oxford, 1955)
2. R.A. Gonzales, J. Eisert, I. Koltracht, M. Neumann, G. Rawitscher, J. Comput. Phys. **134**, 134–149 (1997); R.A. Gonzales, S.-Y. Kang, I. Koltracht, G. Rawitscher, J. Comput. Phys. **153**, 160–202 (1999)
3. G. Rawitscher, I. Koltracht, Comput. Sci. Eng. **7**, 58 (2005)
4. G. Rawitscher, I. Koltracht, Eur. J. Phys. **27**, 1179 (2006)
5. G. Rawitscher, Phys. Rev. C **25**, 2196–2213 (1982)
6. G. Rawitscher, Phys. Rev. E **85**, 026701 (2012)
7. B.D. Shizgal, *Spectral Methods in Chemistry and Physics. Applications to Kinetic Theory and Quantum Mechanics* (Springer, Dordrecht, 2015)

Chapter 11
Sturmian Functions

Abstract Sturmians are a discrete set of eigenfunctions of a Sturm–Liouville equation. They form a complete set in terms of which the solution of numerous equations as a Schrödinger equation can be expanded, both for positive and negative energies. The distinguishing feature of these functions is that they embody the appropriate boundary conditions and already solve a portion of the more complicated Schrödinger equation. This expansion method provides an alternative form to the perturbation theory traditionally used to obtain the solution to a Schrödinger equation in the presence of perturbations. We show how to numerically obtain the Sturmian functions, and how their accuracy can be improved by using spectral methods based on Chebyshev expansions. We also show how these functions serve to obtain a separable representation of a general operator (for example of a non-local potential), and we describe the iterative corrections of the truncation error in a Sturmian expansion.

11.1 Introduction

In this chapter we discuss methods based on Sturmian functions. The advantages of methods that involve such functions are numerous and are clearly described in the introduction of Ref. [1].[1]

In the early days of Quantum Mechanics, before computational methods were in "vogue", the way to find the bound-state energy eigenvalues of a *Schrödinger* equation in the presence of a potential $V(r)$ was by applying a specific iterative method. It consisted in invoking the analytically known eigenvalues $\Lambda_0^{(n)}$ and eigenfunctions $\phi_0^{(n)}$, with $n = 1, 2, \ldots$, in the presence of a potential $V_0(r)$, that could take the form of a harmonic oscillator or a Coulomb potential. These functions form a complete set of eigenfunctions of a Sturm–Liouville differential equation, and hence can be

---

[1]The authors are grateful to G. Gasaneo (Departamento de Fisica, Universidad Nacional del Sur, 8000 Bahía Blanca, Buenos Aires, Argentina) for his contributions to the present chapter.

G. Rawitscher et al., *An Introductory Guide to Computational Methods for the Solution of Physics Problems*,
https://doi.org/10.1007/978-3-319-42703-4_11

used to expand the unknown eigenfunction associated to potential V in an iterative procedure. The method consists in decomposing the potential V into $V_0 + \Delta V$, assuming that ΔV is sufficiently "small", so that the iterations are expected to converge. The method is called "perturbation theory", and can be found in any textbook on Quantum Mechanics. However the convergence of the method is not always assured, as described in detail in the book by Bender and Orszag [2], Chapter 7, and in additional references [3]. An alternative to this form of perturbation theory is to proceed by numerically constructing a set of Sturmian basis functions that need not be known analytically but satisfy the correct boundary conditions, expand the unknown eigenfunction associated to potential V in terms of this basis, and set up a matrix equation for the coefficients and the corresponding eigenvalues. One such method was described in Chap. 9 for the Vibrating Inhomogeneous String. Another example is contained in Chap. 10, where first an approximate set of eigenvalues is found by applying a grid method, and then the values are improved by using an iterative method. Expansion basis functions and their many applications to the solution of physics problems are also discussed very extensively in section 6.7.5 in Ref. [4].

The perturbation methods described above are very useful for the case of bound state eigenvalues, where the functions decay exponentially over large distances. However, in the case of unbound scattering states, for which the asymptotic behavior of the wave function is oscillatory (rather than zero) the commonly used basis functions are usually continuous "plane waves momentum eigenstates" that lead to Dirac Delta functions, and singularities in the corresponding Green's function. Powerful Fourier-Grid methods have been developed in this context [5].

The purpose of the present chapter is to present yet another method that in our case consists in setting up a collection of basis Sturmian functions, that are eigenfunctions of a Sturm–Liouville differential or integral equation, for the purpose of obtaining a positive energy scattering function in the solution of a local or non-local second order Schrödinger equation. Such functions were examined by S. Weinberg and called the "Quasi-Particle" method [6]. We call these functions positive energy Sturmian functions, in contrast to the negative energy Sturmian functions first described by Rotenberg [7], and applied to many physics problems [8–11]. In the present positive energy case the energy of the incident wave is given, but not as an eigenvalue. The method is also suitable for finding eigenfunctions of a more general operator, as will be shown. These auxiliary Sturmian functions all obey the same asymptotic boundary condition, being that they are outgoing complex Hankel functions, with the same energy (or wave number k) as the ingoing plane wave function. They are distorted by a simple auxiliary potential $\bar{V}$ and hence form a very regular and easily checked sequence, but they are all constructed numerically in coordinate (not momentum) space. They are mutually orthogonal with the weight function given by $\bar{V}$. Since $\bar{V}$ vanishes asymptotically, the orthonormality integrals receive contributions only in the finite region where the potential $\bar{V} \neq 0$, and in the region where $\bar{V} \simeq 0$ successive auxiliary Sturmians have an increasing number of oscillations. The eigenvalues $\Lambda_0^{(n)}$ multiply the potential $\bar{V}$ and are complex (real plus imaginary) numbers. The Sturmians can be complex functions, even if the potentials $V(r)$ or $\bar{V}$ are real. The calculations are performed with a spectral expansion into Chebyshev polynomials

[12, 13], with an accuracy expected to be better than 7–8 significant figures, which is desirable for doing atomic physics calculations.

A team of scientists from Argentina recently revived the use of Sturmian functions in two- and three-body problems [14]. They addressed the bound states of the two-electron atoms in both free states, and when confined in a fullerene cage [15] and also included scattering and reaction processes [16]. The double ionization of the He atom by radiation and by electron and heavy ion collisions was studied and produced very successful results [17]. The team also formulated three-body scattering theory in hyperspherical coordinates [18, 19].

This chapter first presents the construction of the auxiliary Sturmian basis functions for positive energies based on the auxiliary potential $\bar{V}$. The method supersedes the one developed previously [20], which considered a square well potential Sturmian basis set in terms of which the desired Sturmians were expanded. Because our method is based on a spectral expansion, it is considerably more precise and flexible, and hence permits a more accurate study of the iteration convergence properties. Next, we present the Sturmians based on a more complicated potential, with a repulsive core. Then, we describe the Sturmian expansion associated to a more general integral operator. Finally, because expansions into Sturmian functions converge slowly [21], an iterative method to make up for an early truncation error of the expansion will be demonstrated.

11.2 Sturmian Functions

The Sturmian functions Φ_s are eigenfunctions of the integral kernel $\mathscr{G}(r, r')\bar{V}(r')$

$$\eta_s \Phi_s(r) = \int_0^\infty \mathscr{G}(r, r')\bar{V}(r')\Phi_s(r')dr', \quad s = 1, 2, 3, \ldots. \quad (11.1)$$

with η_s the eigenvalue, $\mathscr{G}(r, r')$ is the Green's function defined below for a particular Sturmian energy, and $\bar{V}(r')$ is the auxiliary Sturmian potential. The differential Schrödinger equation corresponding to Eq. (11.1) is

$$(d^2/dr^2 + E)\, \Phi_s = \Lambda_s \bar{V}\, \Phi_s, \quad (11.2)$$

with $\Lambda_s = 1/\eta_s$. The Sturmians for positive energies are not square integrable, but they are orthogonal to each other with the weight factor $\bar{V}$ (that is assumed to decrease sufficiently fast with r). The normalization of the Sturmians adopted for most of the present discussion is

$$\left(\Phi_s|\bar{V}\Phi_{s'}\right) = \int_0^\infty \Phi_s(r)\bar{V}(r)\Phi_{s'}(r)\, dr = \eta_s\delta_{s,s'}. \quad (11.3)$$

The result above can be shown by starting from Eq. (11.2), multiplying each side with $\Phi_{s'}$, integrating the result from 0 to ∞ and after an integration by parts and making use of the fact that both Φ_s and $\Phi_{s'}$ obey the same boundary conditions at $r = 0$ (the Φ_s's are zero) and at ∞ (the Φ_s's are both proportional to a Hankel function). Because of the completeness of the Sturmian functions, one has the identity

$$\delta(r - r') = \sum_{s=1}^{\infty} \Phi_s(r) \frac{1}{\left(\Phi_s | \bar{V} \Phi_s\right)} \Phi_s(r') \bar{V}(r'). \qquad (11.4)$$

Generalized Sturmian functions $S_s(r)$, with $s = 1, 2, \ldots$, have also been defined and used extensively by the Argentinian group [22]. Their respective equations

$$\left[d^2/dr^2 - \frac{L(L + 1)}{r^2} - U(r) + E \right] S_s = \beta_s \bar{V} S_s \qquad (11.5)$$

contain an auxiliary potential $U(r)$ together with a short range generating potential $\bar{V}(r)$, where β_s are the eigenvalues. They are also $\bar{V}$ orthogonal

$$\left(\Phi_s | \bar{V} \Phi_{s'}\right) = \int_0^{\infty} S_s(r) \bar{V}(r) S_{s'}(r) \, dr = \gamma_s \delta_{s,s'}. \qquad (11.6)$$

The boundary conditions are the same as the ones imposed on the solution of the Schrödinger equation ψ. The functions S_s, regardless of the value of s, all coincide asymptotically (to within a normalization factor) with ψ, since at large distances $\bar{V}$ is negligible. This is an advantage when U is very long ranged, since the expansion of ψ in terms of the Sturmian functions S_s, $s = 1, 2, 3, \ldots$, focuses only on the radial region where $\bar{V} \neq 0$.

Only the case of short ranged potentials will be treated in the discussion below. The case of long ranged potentials, such as the Coulomb potential, has been considered by means of the generalized Sturmian functions in Ref. [22].

11.2.1 Positive Energies

For the case that the orbital angular momentum L is zero, the positive energy Green's function $\mathscr{G}(r, r')$ in Eq. (11.1) is given by

$$\mathscr{G}(r, r') = -\frac{1}{k} F(r_<) \times H(r_>), \qquad (11.7)$$

where $(r, r') = (r_<, r_>)$ if $r \leq r'$ and $(r, r') = (r_>, r_<)$ if $r \geq r'$. Here

$$F(r) = \sin(kr); \quad H(r) = \cos(kr) + i \sin(kr) \qquad (11.8)$$

and k is the wave number, in terms of which the energy $E = k^2$ is defined. For positive energies and short range potentials the Sturmians obey the boundary conditions $\Phi_s(r \to 0) = 0$; $\Phi_s(r \to \infty) = \ell_s H(r)$, where the constant ℓ_s is determined by the normalization of the Sturmian function. The asymptotic form of ψ is

$$\psi(r \to \infty) = F(r) + S\,H(r), \tag{11.9}$$

with

$$S = -\frac{1}{k}\int_0^\infty F(r')V(r')\psi(r')\,dr'. \tag{11.10}$$

A generalization to angular momenta $L > 0$ can be easily accomplished [13].

The numerical calculation of the eigenvalues and eigenfunctions proceeds along the same lines as explained in Chap. 10. First an approximate spectrum of the eigenvalues is obtained by means of the eigenvalues of the matrix that is obtained by means of the Chebyshev spectral expansion of the operator

$$\int_0^\infty \mathscr{G}(r,r')\bar{V}(r')\Phi_s(r')dr'$$

that appeared in Eq. (11.1), in a similar way to what was done in Chap. 6. The next step is to perform an iterative improvement of the eigenvalues through the iterative procedure based on Hartree's method, also demonstrated in Chap. 10, which improves the accuracy to $1 : 10^{-11}$. This procedure supersedes a method previously described in Ref. [20].

One can understand intuitively the properties of the Φ's as follows [20]. By comparing the Eq. (11.1) with Eq. (11.2) one sees that $\Lambda_s = 1/\eta_s$. As the index s increases, the corresponding values of Λ_s increase, and hence the potential $\Lambda_s\bar{V}$ in Eq. (11.2) increases in magnitude. If $\bar{V}$ is real and attractive and the real part of Λ_s is positive, then the real part of $\Lambda_s\bar{V}$ becomes more attractive, and the corresponding eigenfunction Φ_s becomes more oscillatory inside of the attractive region of the well. Therefore, from one s to the subsequent $s + 1$ the eigenfunction acquires one more node inside of the well. According to flux considerations the imaginary part of $\Lambda_s\bar{V}$ has to be positive, meaning that the well has to be emissive [20] so as to correspond to the outgoing nature of the asymptotic function H. This is exactly the opposite of the case of an optical potential (that absorbs flux), where the imaginary part is negative.

These properties will be verified in the numerical illustrations below for three examples with $k = 0.5\,\mathrm{fm}^{-1}$ in each case. In the first example the potential $\bar{V}$ is given by

$$V_S = 6\,\exp(-0.3\,r)\,\big[\exp(-0.3\,r) - 2\big]; \tag{11.11}$$

in the next example the potential is of the Woods Saxon form

$$V_{WS} = V_0/\{1 - \exp[(r - R)/a]\}, \tag{11.12}$$

with $V_0 = -5\,\text{fm}^{-2}$, $R = 15\,\text{fm}$ and $a = 0.5\,\text{fm}$. Both potentials V_S and V_{WS} have no repulsive core. For the third example, $\bar{V}$ is of the Morse type with a repulsive core near the origin, given by

$$V_P(r) = 6\exp(-0.3\,r + 1.2) \times \left[\exp(-0.3\,r + 1.2) - 2\right]. \qquad (11.13)$$

The number 6 is given in units of fm^{-2}, the number 0.3 is in units of fm^{-1}, r is given in units of fm, and the other constants have no dimensions. These Sturmian potentials illustrated in Fig. 11.1 are in units of inverse length squared, and were transformed from their energy units by multiplication with the well known factor $2\mu/\hbar^2$. The energy E is related to the wave number k according to $E = k^2$.

The Λ-spectrum for potential V_S is shown in Fig. 11.2. As expected, the imaginary parts of Λ_s are slightly negative and the real parts are increasingly positive, with the result that $\Lambda_s V_S$ becomes unceasingly attractive as s increases.

The corresponding Sturmian eigenfunctions acquire increasingly more nodes in the attractive region of the potential, as is illustrated in Figs. 11.3 and 11.4.

For future benchmark purposes, the eigenvalues of Eq. (11.1) are given in the Table 11.1 for the Sturmian potential V_S defined in Eq. (11.13), with a wave number $k = 0.5\,\text{fm}^{-1}$.

When the potential $\bar{V}$ has a repulsive core, as is the case for potential V_P, Eq. (11.13), the Sturmians change in character, because some of the real parts of Λ_s are negative, as is illustrated in Fig. 11.5 for the spectrum of $\Lambda_s = 1/\eta_s$.

This can be understood as follows: because potential V_P has both a repulsive and an attractive part, the eigenvalues fall into two categories. In category I the eigenvalues Λ have a positive real part and a negative imaginary part, and the corresponding eigenfunctions are large mainly in the attractive regions of the potential well. Examples are given in Figs. 11.6 and 11.7.

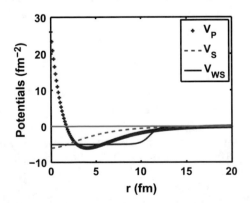

Fig. 11.1 Three Sturmian potentials $\bar{V}$ as a function of radial distance, given by Eqs. (11.13), (11.11) and (11.12). V_S (dashed green line) and V_{WS} (solid red line) are introduced because they have different ranges, while the blue line with crosses represents a more realistic potential

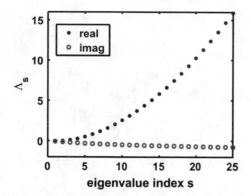

Fig. 11.2 The spectrum of the Sturmian eigenvalues Λ_s for the potential V_S defined in Eq. (11.11) and illustrated in Fig. 11.1. The wave number is $k = 0.5\,\text{fm}^{-1}$. The larger the real part of the eigenvalue, the deeper the real part of potential $\Lambda \bar{V}$ becomes, and consequently the corresponding Sturmian eigenfunction has more minima. The text explains why these eigenvalues have to be complex

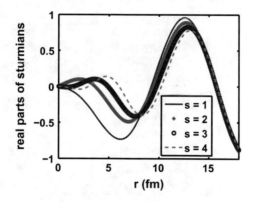

Fig. 11.3 Sturmian eigenfunctions for the potential V_S defined in Eq. (11.11) for a wave number $k = 0.5\,\text{fm}^{-1}$. They are normalized such that asymptotically they all approach the outgoing Hankel function $H(r) = cos(kr) + i\ sin(kr)$. As the index s increases, the corresponding Sturmian function acquires one more node in the region where the potential is large

In category II the real parts of Λ are negative thus transforming the repulsive piece of the potential near the origin into an attractive well, and transforming the formerly attractive valley into a repulsive barrier. The resulting potential is similar to the one examined in Chap. 6, where resonances occurred for certain energies. Examples of the corresponding Sturmian indices are $s = 5, 7, 10, 15, \ldots$. One of these functions for $s = 5$ is shown in Fig. 11.8.

The Sturmian for $s = 10$ is similar to that for $s = 5$, because it is also large near the origin (with an amplitude of $\simeq 10^9$) and has a node near $r = 1$. The functions for $s = 5$ and 10 are "resonant" in the radial region $r \in [0, 4]$, while the one for $s = 7$ is non-resonant. At a larger energy the effect of the barrier for the functions

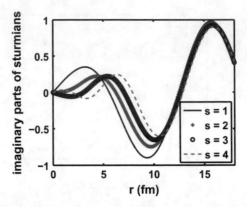

Fig. 11.4 Plot of the same results as Fig. 11.3 for the imaginary part of the Sturmian functions. The imaginary part of the Sturmian wave functions have more nodes than the real parts for the same value of the index s. This is particularly evident for the blue curve ($s = 1$)

Table 11.1 Eigenvalues for Sturmian potential V_S for $k = 0.5\,\mathrm{fm}^{-1}$

	Real part of Λ_s	Imag. part of Λ_s
1	−0.03297806784	−0.05633093256
2	0.04181033607	−0.12298817955
3	−0.02140651197	−0.23641901361
4	0.21055103262	−0.15935376881
5	0.43743611684	−0.17810671020
6	0.71976101832	−0.19289033454
7	1.05649701588	−0.20516494913
8	1.44649394898	−0.21538251834
9	1.88887575723	−0.22390738772
10	2.38303461225	−0.23106484062
11	2.92855765503	−0.23712639599
12	3.52516147256	−0.24230800967
13	4.17264661513	−0.24677772914
14	4.87086853262	−0.25066566005
15	5.61971940360	−0.25407296058
16	6.41911669954	−0.25707898962
17	7.26899583793	−0.25974669404
18	8.16930533353	−0.26212658464
19	9.12000350403	−0.26425965551
20	10.1210561674	−0.26617953514
21	11.1724349888	−0.26791408364
22	12.2741162656	−0.26948659119

Fig. 11.5 The spectrum of the eigenvalues Λ_s for the potential V_P defined in Eq. (11.13). This potential has a repulsive core. The wave number is $k = 0.5\,\text{fm}^{-1}$. The real parts of Λ can have negative or positive values, in contrast to the case with only positive values shown in Fig. 11.2

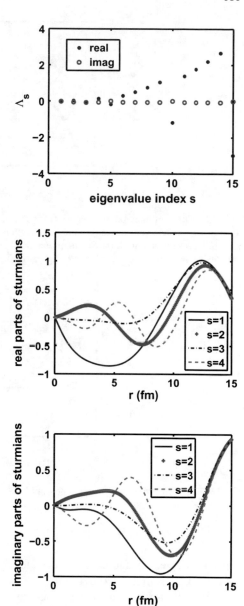

Fig. 11.6 Real parts of Sturmian functions Φ_s for the potential V_P, for $k = 0.5\,\text{fm}^{-1}$. This potential defined in Eq. (11.13) has a repulsive core. For this reason the shape of these functions is much less predictable that for the case where the potential is monotonic. The result for the anomalous $s = 5$ function is shown in a separate figure

Fig. 11.7 Same as Fig. 11.6 for the imaginary parts of the Sturmian functions

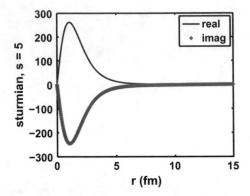

Fig. 11.8 Sturmian function in category II for $s = 5$, potential V_P and $k = 0.5$. This result illustrates a shape resonance, as seen by the large magnitude of these functions in the barrier region. For other cases in category II the Sturmian function can be very small in the region of the potential, and outside the region they become of order unity. The real part is illustrated by the blue solid line

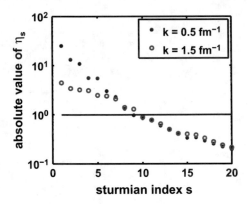

Fig. 11.9 Absolute value of η_s as a function of the Sturmian index s for two values of the wave number k, both for a positive energy. The eigenvalues η_s are defined in Eq. (11.1) or (3.8), with $\bar{V} = V_P$

of class II decreases, and the absolute value of the eigenvalues $\eta_s = 1/\Lambda_s$ decreases for $s < 10$, as is illustrated in Fig. 11.9.

The functions for which $|\eta_s| < 1$ play an important role for the iterative correction of the truncation errors, as is shown in Appendix B of Ref. [23]. Further, in the expansion of a wave function in terms of Sturmians which are themselves eigenfunctions of the integral operator, a dominator $(1 - \eta_s)$ is likely to appear in the expansion of the wave function. With regards to the values of s for which $real(\eta_s) \simeq 1$, and $imag(\eta_s) \simeq 0$, the corresponding Sturmians make a resonant contribution to that expansion. The real and imaginary parts of some of the η_s are illustrated in Fig. 11.10. This illustration takes the form of an Argand diagram for three values of k, which

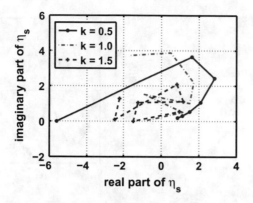

Fig. 11.10 Argand diagram of η_s, for the potential V_P for $s = 4, 5, \ldots, 10$, for three values of k (in units of fm^{-1}). The green dash dot line indicates the results for $k = 1.0$. The sturmian energy is positive. The $s = 7$ points for all three wave numbers are close to the resonance condition, for which the real parts of $\eta_s \simeq 1$, and the imaginary parts of $\eta_s \simeq 0$

shows that for $s = 7$ the values of η_s for the three values of k satisfy the resonance criterion by lying close to unity.

11.2.2 Negative Energies

For negative energies $E = -\kappa^2$, the Green's function $\bar{\mathcal{G}}_0$ is

$$\bar{\mathcal{G}}_0(r, r') = -\frac{1}{\kappa}\bar{F}(r_<) \times \bar{H}(r_>), \qquad (11.14)$$

where again $(r, r') = (r_<, r_>)$ if $r \le r'$ and $(r, r') = (r_>, r_<)$ if $r \ge r'$, and where

$$\bar{F}(r) = \sinh(\kappa r); \quad \bar{H}(r) = \exp(-\kappa r). \qquad (11.15)$$

The case of the Woods–Saxon potential is shown in Eq. (11.12). For a negative energy wave number $\kappa = 0.3\,\mathrm{fm}^{-1}$ the first four Sturmian functions are displayed in Fig. 11.11, while the eigenvalues $\Lambda_s = 1/\eta_s$ are shown in Fig. 11.12.

The normalization integral $\langle \bar{\Phi}_s \bar{V} \bar{\Phi}_s \rangle = \langle \bar{\Phi}_s^2 \bar{V} \rangle$ in Eq. (11.3) is negative (since the potential $\bar{V}$ is negative definite), hence the Sturmians $\bar{\Phi}_s$ are purely imaginary, and are displayed in Fig. 11.11.

Fig. 11.11 Negative energy
Sturmians $\tilde{\Phi}_s$, for the
potential V_{WS} defined in
Eq. (11.12), with
$\kappa = 0.3\,\text{fm}^{-1}$. In a similar
way for the positive energy
case, although less clearly
visible (Fig. 11.3), the larger
the index s, the more nodes
occur in the wave function

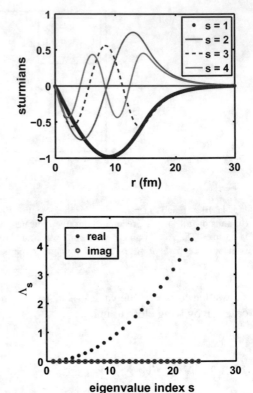

Fig. 11.12 Eigenvalues Λ_s
for the negative energy
Sturmians with
$\kappa = 0.3\,\text{fm}^{-1}$, for the
potential V_{WS} defined in
Eq. (11.12). Only the real
parts of the eigenvalues Λ_S
are different from zero.
Some of the corresponding
Sturmian functions are
illustrated in Fig. 11.11

11.3 Solution of the Schrödinger Equation for Positive Energies

11.3.1 Local Potential V(r)

Of course the easiest path to obtain the scattering function is to use a conventional
method (finite element or finite difference). However, when we are faced with cases
of non-local potentials solved with a Sturmian expansion, it is instructive to first
consider the local case.

The equation to be solved (again for a radial wave function with $L = 0$) is

$$\left(\frac{d^2}{dr^2} + k^2\right)\psi(r) = V(r)\psi(r). \qquad (11.16)$$

In preparation for the Sturmian expansion, it is convenient to separate the solution
into an incident part $F(r)$ and a scattering part $\chi(r)$

$$\psi = F + \chi, \tag{11.17}$$

and remembering that $(\frac{d^2}{dr^2} + k^2)F(r) = 0$, Eq. (11.16) can be written as

$$\left(\frac{d^2}{dr^2} + k^2 - V\right)\chi(r) = VF(r). \tag{11.18}$$

Since the asymptotic limit of χ is an outgoing Hankel function, it is appropriate to expand χ in terms of the auxiliary Sturmian functions $\Phi_s(r)$, with expansion coefficients c_s to be determined, since the Φ_s's obey the same asymptotic boundary conditions as

$$\chi_{(N_s)}(r) = \sum_{s=1}^{N_s} c_s \Phi_s(r). \tag{11.19}$$

The subscript (N_s) is to indicate that the expansion has been truncated at the upper limit N_s. Therefore, $\chi_{(N_s)}$ is only an approximation to χ, whose error is corrected iteratively in Sect. 11.5. By multiplying both sides of Eq. (11.18) with $\Phi_{s'}$, integrating from $r = 0$ to $r = \infty$, and making use of the normalization implicit in Eq. (11.3), together with $\eta_s \Lambda_s = 1$, one finally obtains the matrix equation for the expansion coefficients c_s

$$\sum_{s=1}^{N_s} M_{s's} c_s = (\Phi_{s'}|VF), \tag{11.20}$$

where the symmetric matrix $M_{s's}$ is given by

$$M_{s's} = \delta_{s's} - (\Phi_{s'}|V\Phi_s). \tag{11.21}$$

It is useful to observe that:

(a) If $V = \bar{V}$, then $M_{s's}$ becomes diagonal and $c_s = (\Phi_{s'}|VF)/(1 - \eta_s)$. The presence of this denominator shows that the Sturmians whose (complex) eigenvalues have a real part close to unity contribute the most to the outgoing scattering wave. This is the case for the $s = 7$ Sturmians illustrated in Fig. 11.10 for all three values of the wave numbers k. This observation shows that knowledge of the spectrum of Sturmians can be useful in identifying the location of resonances [11].

(b) If $V = \bar{V} + \Delta V$, where ΔV is small compared to $\bar{V}$, the properties of case (a) remain approximately valid. For this case the matrix M becomes $M_{s's} = \delta_{s's}(1 - \eta_s) - (\Phi_{s'}|(\Delta V)\Phi_s)$, and the off-diagonal part of this matrix can be included iteratively.

11.3.2 Non-local Potential V

For this case the expansion into Sturmian functions becomes very useful, since these functions are "global", and the finite element or finite difference methods become less useful. One exception is the solution of the integral equation for ψ, as will be shown further on.

We introduce the integration kernel $K(r, r')$, and replace $V\psi$ by

$$V\psi \rightarrow \int_0^\infty K(r, r')\psi(r')dr'$$

in Eq. (11.16). The derivations that led to Eqs. (11.20) and (11.21) are still valid, and the only necessary change is to replace $(\Phi_{s'}|V\Phi_s)$ and $(\Phi_{s'}|VF)$ by

$$(\Phi_{s'}|V\Phi_s) \rightarrow \int\int \Phi_{s'}(r')K(r', r'')\Phi_s(r'')dr'dr'', \tag{11.22}$$

$$(\Phi_{s'}|VF) \rightarrow \int\int \Phi_{s'}(r')K(r', r'')F(r'')dr'dr''. \tag{11.23}$$

These double integrals can be carried out since all the functions in the integrand are known. They reduce to sums of products of single integrals if K can be approximated by a separable expression.

11.3.3 Solution of the Lippmann–Schwinger Integral Equation

The equation to be solved is

$$\psi(r) = F(r) + \int_0^\infty \mathscr{G}(r, r')V(r')\psi(r')dr'. \tag{11.24}$$

However, by making use in the Galerkin formalism of Eqs. (11.1) and (11.3) , one is led again to the same Eqs. (11.20) and (11.21) for the coefficients c_s of the expansion (11.19). The argument is as follows. Starting from Eq. (11.24), multiplying each side with $\Phi_{s'}(r)\bar{V}(r)$ and integrating over r from 0 to ∞, one finds

$$\sum_{s=1}^{N_s} \langle \Phi_{s'}\bar{V}\Phi_s \rangle c_s = \left(\Phi_{s'}\bar{V}\mathscr{G}VF \right) + \sum_{s'=1}^{N_s} \left(\Phi_{s'}\bar{V}\mathscr{G}V\Phi_s \right) c_s. \tag{11.25}$$

If one then makes use of $\left(\Phi_{s'}\bar{V}\mathscr{G}VF\right) = \eta_{s'}\left(\Phi_{s'}VF\right)$, and $\left(\Phi_{s'}\bar{V}\mathscr{G}V\Phi_s\right) = \eta_{s'}\left(\Phi_{s'}V\Phi_s\right)$ in Eq. (11.25), and cancels $\eta_{s'}$ on both sides, then Eqs (11.20) and (11.21) are recovered, regardless of whether V is local or non-local.

11.4   Separable Expansion of a General Integral Operator

The general one-dimensional integral equation to be solved for ψ is

$$\psi(r) = F(r) + \int_0^\infty \mathscr{O}(r, r'')\psi(r'')\, dr'', \tag{11.26}$$

where F is the driving term and $\mathscr{O}$ is a general integration kernel, both assumed to be known. The shorthand form of the above equation is

$$\psi = F + \mathscr{O}\psi. \tag{11.27}$$

In order to obtain a separable representation of the operator $\mathscr{O}$ two methods are possible:

(1) one can rewrite $\int_0^\infty \mathscr{O}(r, r'')\psi(r'')\, dr''$ as

$$\int_0^\infty \int_0^\infty dr'\delta(r - r')dr'\mathscr{O}(r', r'')\psi(r'')\, dr,$$

and replace $\delta(r - r')$ by Eq. (11.4). If the sum in Eq. (11.4) is truncated at an upper limit N_s one obtains a finite rank expansion of the delta function

$$\delta_{N_s}(r, r') = \sum_{s=1}^{N_s} \Phi_s(r)\frac{1}{\langle\Phi_s\bar{V}\Phi_s\rangle}\Phi_s(r')\bar{V}(r'). \tag{11.28}$$

Please note that $\bar{V}$ appears as a function of the variable of integration, r'. The corresponding finite rank representation of the operator $\mathscr{O}$ is

$$\mathscr{O}_N^{(1)}(r, r'') = \sum_{s=1}^{N_s} \Phi_s(r)\frac{1}{\langle\Phi_s\bar{V}\Phi_s\rangle}O_s^{(1)}(r''), \tag{11.29}$$

where

$$O_s^{(1)}(r'') = \int_0^\infty \Phi_s(r')\bar{V}(r')\mathscr{O}(r', r'')dr'. \tag{11.30}$$

Hence a separable representation of $\mathscr{O}\psi$ by method (1) is

$$\mathscr{O}\psi \rightarrow \sum_{s=1}^{N_s} \Phi_s(r) \frac{1}{\left(\Phi_s \bar{V} \Phi_s\right)} \int_0^\infty O_s^{(1)}(r'')\psi(r'')dr''. \tag{11.31}$$

For example, if $\mathscr{O}(r', r'') = \mathscr{G}(r', r'')V(r'')$, then $O_s^{(1)}(r', r'') = \eta_{ss}(r'')V(r'')$ which follows from (11.30) in view of Eqs. (11.1) and (11.3). As a result one obtains a separable representation of the Green's function

$$\mathscr{G}_{N_s}(r, r') = \sum_{s=1}^{N_s} \Phi_s(r)\Phi_s(r'). \tag{11.32}$$

This result, when inserted into Eq. (11.26), and using the Galerkin formalism, leads again to Eqs. (11.20) and (11.21) for the expansion coefficients in Eq. (11.19).
(2) On the other hand, if one rewrites $\int_0^\infty \mathscr{O}(r, r')\psi(r')\,dr'$ as

$$\int_0^\infty \int_0^\infty dr' \mathscr{O}(r, r')\delta(r'' - r')\psi(r'')\,dr'',$$

and replaces the delta function by the separable representation

$$\delta_{N_s}(r'' - r') = \sum_{s=1}^{N_s} \Phi_s(r') \frac{1}{\left(\Phi_s \bar{V} \Phi_s\right)} \Phi_s(r'')\bar{V}(r''),$$

one obtains the second form for the separable representation of the operator $\mathscr{O}$

$$\mathscr{O}_N^{(2)}(r, r') = \sum_{s=1}^{N_s} O_s^{(2)}(r) \frac{1}{\left(\Phi_s \bar{V} \Phi_s\right)} \Phi_s(r'')\bar{V}(r''), \tag{11.33}$$

with

$$O_s^{(2)}(r) = \int_0^\infty \mathscr{O}(r, r')\Phi_s(r')dr'. \tag{11.34}$$

In this case the separable representation of $\mathscr{O}\psi$ is

$$\mathscr{O}\psi \rightarrow \sum_{s=1}^{N_s} O_s^{(2)}(r) \frac{1}{\left(\Phi_s \bar{V} \Phi_s\right)} \int_0^\infty \Phi_s(r'')\bar{V}(r'')\psi(r'')\,dr''. \tag{11.35}$$

For example, in the case that $\mathscr{O} = d/dr$, Eq. (11.35) becomes

$$(d/dr)_N = \sum_{s=1}^{N} |\Phi'_s\rangle \frac{1}{\left(\Phi_s \bar{V} \Phi_s\right)} \langle \Phi_s \bar{V}, \tag{11.36}$$

where $\Phi'_s = d\Phi_s/dr$. This equation provides a finite rank integral approximation to the derivative operator. In Eq. (11.36) the symbol $|\Phi'_s\rangle$ is shorthand for a function, while $\langle \Phi_s \bar{V}$ represents an integral operator to be applied to the function standing on it's right. As an application of Eq. (11.36), when

$$f_t(r) = \frac{1}{\bar{a}} \exp((r - \bar{R})/\bar{a})/[1 + \exp((r - \bar{R})/\bar{a})]^2, \tag{11.37}$$

then the finite rank approximation to df_t/dr is

$$f_t'^{(N)}(r) = \sum_{s=1}^{N} |\Phi'_s\rangle \frac{1}{\left(\Phi_s \bar{V} \Phi_s\right)} \left(\Phi_s \bar{V} f_t\right). \tag{11.38}$$

In Eq. (11.38) the symbol $\left(\Phi_s \bar{V} f_t\right)$ is no longer an operator but is a number given by a definite integral. As an example, when $\bar{R} = 3.5\,\text{fm}$ and $\bar{a} = 0.5\,\text{fm}$ in Eq. (11.37), and using $N = 24$ negative energy Sturmians Φ_s one obtains a numerical result for $f_t'^{(N)}(r)$ that is accurate to somewhat better than $1 : 10^{-2}$. Both f_t and $f_t'^{(N)}(r)$ are illustrated in Fig. 11.13, where they are labeled as "input" and "output", respectively. The value of $\kappa = \sqrt{-E}$ is $0.3\,\text{fm}^{-1}$, and the Sturmian potential is V_{WS}, as defined in Eq. (11.12).

If the same calculation of the derivative is done by using the Chebyshev derivative matrix, with 21 Chebyshev basis functions, one obtains the same type of accuracy as that shown in Fig. 11.13. The derivative matrix is listed in Appendix B, where it is denoted as "test_2_deriv2". If the number of Chebyshev basis functions is increased from 21 to 31, the accuracy improves by more than one order of magnitude.

Fig. 11.13 A numerical application for the representation of a derivative, Eq. (11.38), applied to the function $f_t(r)$ defined in Eq. (11.37), and denoted as "input". The approximation $f_t''^{(N)}(r)$ is denoted as "output". With 24 negative energy Sturmians the error for $f_t''^{(24)}(r)$ is less than $1 : 10^{-2}$

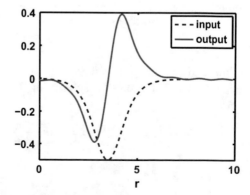

11.5   Iterative Correction of the Truncation Error

11.5.1   Method $\mathscr{S}_1$

The iterative solution of Eq. (11.26) is achieved by first approximating the operator $\mathscr{O}$ by a separable representation $\mathscr{O}_N$ of rank N, defining the remainder $\Delta_N^{(1)}$ as

$$\Delta_N^{(1)} = \mathscr{O} - \mathscr{O}_N, \tag{11.39}$$

and then iterating on the remainder. If the norm of $\Delta_N^{(1)}$ is less than unity, the iterations should converge. Since the numerical complexity of performing iterations is less than the complexity of solving a linear equation with a matrix of large dimension, this method can be computationally advantageous, and furthermore the exact eigenfunctions of the operator $\mathscr{O}$ need not be known. Therefore, the aim of this section is to develop the iteration scheme and observe the rate of convergence via some numerical examples.

The approximate discretization of the kernel $\mathscr{O}$ into a representation of rank N is given by Eq. (11.35), which can be written in the form

$$\mathscr{O}_N(r, r') = \sum_{s=1}^{N} |\mathscr{O} \, \Phi_s\rangle \frac{1}{(\Phi_s \bar{V} \Phi_s)} \langle \Phi_s \bar{V}. \tag{11.40}$$

Here the symbol "$\rangle$" denotes that the function $|\mathscr{O} \, \Phi_s\rangle$ is evaluated at position r, and "$\langle$" denotes that the quantity to the right of it is evaluated at r' and integrated over r'. The quantity $\langle \Phi_s \bar{V} \Phi_s \rangle = \eta_s$ denotes the integral $(\Phi_s \bar{V} \Phi_{s'}) = \int_0^\infty \Phi_s(r) \bar{V}(r) \Phi_{s'}(r) dr = \delta_{s\,s'} \eta_s$, and $\langle \Phi_s$ denotes an integral to be performed later, but where Φ_s is *not* the complex conjugate of Φ_s, and $\bar{V}(r)$ is the potential used in the definition of the Sturmians.

The iterative procedure of solving Eq. (11.27) denoted as $\mathscr{S}_1$ consists in first obtaining a function $\mathscr{F}^{(1)}(r)$ that is the solution of

$$\mathscr{F}^{(1)} = F + \mathscr{O}_N \mathscr{F}^{(1)}, \tag{11.41}$$

followed by an iteration on the remainder $\Delta_N^{(1)}$. Because of the separable nature of $\mathscr{O}_N$, given by Eq. (11.33), the solution of (11.41) is algebraic, and is given by

$$\mathscr{F}^{(1)}(r) = F(r) + \sum_{s=1}^{N} c_s^{(1)} |\mathscr{O} \, \Phi_s\rangle_r, \tag{11.42}$$

where the $c_s^{(1)}$, $s = 1, 2, \ldots, N$ are the solutions of the matrix equation

$$\sum_{s'=1}^{N} \left(\delta_{s,s'} - \frac{1}{(\Phi_s \bar{V} \Phi_s)} \langle \Phi_s \bar{V} \mathcal{O} \Phi_{s'} \rangle \right) c_{s'}^{(1)} = \frac{1}{(\Phi_s \bar{V} \Phi_s)} \langle \Phi_s \bar{V} F \rangle. \tag{11.43}$$

The derivation of Eq. (11.43) can proceed by setting

$$\mathcal{F}^{(1)}(r) = \sum_{s=1}^{N} c_s^{(1)} \Phi_s(r) \tag{11.44}$$

and inserting this expression into Eq. (11.41). By multiplying each side, term by term with $\Phi_{\bar{s}}(r)\bar{V}(r)$ and integrating over r, making use of Eq. (11.3), and summing over all $\bar{s}$, one obtains Eq. (11.43).

An interesting application of Eqs. (11.42) and (11.43) is for the solution of a Schrödinger equation that contains a general nonlocal potential, such as the one given by Perey and Buck [24]. The function $\mathcal{F}^{(1)}$ obtained with 10 Sturmians gives an approximation to ψ that has an error of less than 0.1% [25].

The iterations on the remainder $\Delta_N^{(1)}$ proceed according to

$$\psi = \mathcal{F}^{(1)} + \chi_2^{(1)} + \chi_3^{(1)} + \cdots, \tag{11.45}$$

where the $\chi_n^{(1)}$ are calculated iteratively according to

$$(1 - \mathcal{O}_N)\chi_{n+1}^{(1)} = \Delta_N^{(1)}\chi_n^{(1)}, \quad n = 1, 2, \ldots, \tag{11.46}$$

with $\chi_1^{(1)} = \mathcal{F}^{(1)}$. Equation (11.46) obeyed by $\chi_{n+1}^{(1)}$ is similar to Eq. (11.44) for $\chi^{(1)}$, with the driving term $\mathcal{O}_N F$ replaced $(\mathcal{O} - \mathcal{O}_N)\chi_n^{(1)}$. Hence the solution can be achieved in a similar way and is also algebraic. Numerical examples are given in the next section.

11.5.2 Method $\mathcal{S}_2$

It is also found that instead of solving Eq. (11.27), the once iterated form $\psi = F + \mathcal{O}(F+\mathcal{O}\psi)$

$$\psi = F + \mathcal{O}F + \mathcal{O}^2\psi \tag{11.47}$$

is to be solved for ψ, then the iterations called $\mathcal{S}_2$ will converge faster, as will be verified in the context of the numerical examples in Sect. 11.6, and as is formally demonstrated in Appendix B of Ref. [23].

In this iteration method Eq. (11.41) F is replaced by $F + \mathcal{O}F$, $\mathcal{O}_N$ is replaced by $(\mathcal{O}_N)^2$, and the residue $\Delta_N^{(2)}$ is defined as $\Delta_N^{(2)} = \mathcal{O}^2 - (\mathcal{O}_N)^2$ (11.41). In addition F

is replaced by $F + \mathcal{O}F$, $\mathcal{O}_N$ is replaced by $(\mathcal{O}_N)^2$, the residue $\Delta_N^{(2)}$ is defined as

$$\Delta_N^{(2)} = \mathcal{O}^2 - (\mathcal{O}_N)^2 \tag{11.48}$$

and $\mathcal{F}^{(1)}$ is replaced by $\mathcal{F}^{(2)}$, which is the solution of

$$\mathcal{F}^{(2)} = F + \mathcal{O}F + (\mathcal{O}_N)^2 \mathcal{F}^{(2)}. \tag{11.49}$$

The equation for $\mathcal{F}^{(2)}$ can again be solved algebraically. The terms $\chi_n^{(2)}$ required for the subsequent iterations,

$$\psi = \mathcal{F}^{(2)} + \chi_2^{(2)} + \chi_3^{(2)} + \chi_4^{(2)} + \cdots \tag{11.50}$$

are obtained by solving

$$\chi_{n+1}^{(2)} = (\mathcal{O}_N)^2 \chi_{n+1}^{(2)} + \Delta_N^{(2)} \chi_n^{(2)}, \quad n = 1, 2, \ldots. \tag{11.51}$$

Further details are given in Ref. [23].

11.6   Numerical Examples

The main purpose of this section is to investigate the rate of convergence of the iterative solution of Eq. (11.27), by expanding the operator $\mathcal{O}$ into Sturmians that are not eigenfunctions of $\mathcal{O}$. All the numerical examples are for the case that $\mathcal{O} = \mathcal{G} V$.

In this example the iterations are performed with method $\mathcal{S}_2$, the scattering potential is given by V_P, Eq. (11.13), and by using one of the three potentials as generating the Sturmian functions. The objective is to expand the solution ψ of a potential that has an attractive valley and a repulsive core in terms of Sturmians that are based on monotonic potentials without a repulsive core, and then examining which of the methods converge faster.

The convergence of the iterations is illustrated in Fig. 11.14, where the results with the Sturmians based on potential V_S and V_{WS} are denoted as $S2$ and $WS2$.

The range of potential V_{WS} is comparable to the range of the scattering potential V_P, and consequently the convergence of the iterations is faster if the Sturmians are based on a potential $\bar{V}$ whose range is comparable to the range of the scattering potential. The fastest convergence (labeled "P") is obtained with the Sturmians based on potential V_P, but these Sturmians are as difficult to calculate as the scattering function itself.

As mentioned above, the points labeled "P-Sturmians" in Fig. 11.14 are obtained using the Sturmians for the scattering potential, $\bar{V} = V_P$, and hence the Sturmian functions Φ_s defined by Eq. (11.1) are the same, to within a normalization constant, as the eigenfunctions of the operator $\mathcal{O} = \mathcal{G}_0 V_P$. This method was introduced by

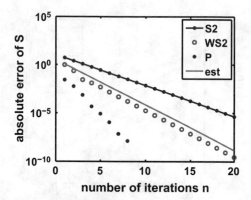

Fig. 11.14 The rate of convergence of the asymptotic limit S of the wave function to the "exact" one, defined by Eqs. (3.6) and (11.10), as a function of the number of iterations n. The Sturmian functions S and WS are obtained with potentials V_S and V_{WS}, respectively, which have no repulsive cores. Iteration method $\mathscr{S}_2$ was used for results V_S and V_{WS} and method $\mathscr{S}_1$ was used for potential V_P. (The latter is identical to the potential used to calculate the scattering wave function). The solid line, denoted as "est" (for estimate) is given by $3.48 \times (0.34)^n$. The wave number is $k = 0.5\,\mathrm{fm}^{-1}$, the number of WS Sturmians is 31

S. Weinberg [6], and is denoted as the quasi-particle method (Q-P). In this case the matrix $M_{s's}$, Eq. (11.21), becomes diagonal, $\delta_{s,s'}\eta_s$, and many of the equations simplify. Because these (Q-P) Sturmians take into account ab initio the repulsive core and attractive valley of the scattering potential, it is not surprising that this method converges fastest. This figure shows that if the eigenfunctions of the operator $\mathscr{O}$ were available, then the Q-P method would be the method of choice. However, the present investigation for a general kernel $\mathscr{O}$ assumes that the (Q-P) Sturmians are not known.

The convergence of method $\mathscr{S}_1$, based on Eqs. (11.41)–(11.46), is considerably slower than for $\mathscr{S}_2$, as illustrated in Fig. 11.15.

The points labeled B and $B2$ are obtained by using methods S_1 and S_2, respectively using the Sturmians based on potential $\bar{V} = V_B$, defined in the equation

$$V_B(r) = V_P(r)\,[1 - exp(-(r/0.5)^2)]. \tag{11.52}$$

This Sturmian potential is identical to potential V_P at large distances, but its repulsive core near the origin is changed into a small repulsive barrier that decreases to zero as $r \to 0$. Points $S2$ are obtained with method $\mathscr{S}_2$ using the Sturmians based on potential $\bar{V} = V_S$. The fact that both Sturmians B and S give nearly indistinguishable results for the iteration, as shown by the symbols $+$ and by the solid line, respectively, shows that the behavior of the Sturmians near the origin does not significantly affect the results, provided that there is no repulsive core in the Sturmian potentials. The open circles in Fig. 11.15, labeled as $Gr - B$, were obtained with a Green's function iteration method, in which potential V_p is divided into $V_B + (V_P - V_B)$. The (L–S) equation with potential V_B is solved exactly (not using the algebraic Eq. (11.42)) to

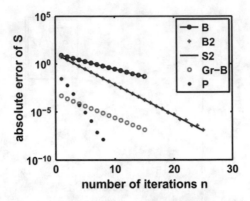

Fig. 11.15 The convergence of the iterations $n = 1, 2, \ldots$ as measured by the error of the asymptotic value S of the wave function ψ, Eq. (11.10). The results labeled B and $B2$ are obtained with Sturmian potential $\bar{V} = V_B$ for methods $\mathscr{S}_1$ and $\mathscr{S}_2$, respectively. The solid line labeled as $S2$ is obtained with method $\mathscr{S}_2$ for the Sturmian potential V_S. The result P is obtained with the original Q-P method, with Sturmians obtained for potential V_P. The open circles, labeled $Gr - B$, are obtained with Green's function iterations, based on $V_P - V_B$, described in the text. The results obtained by using the Sturmian functions with potential $\bar{V} = V_P$ give the smallest error, while the Green's function method (green open circles) gives the next best accuracy

produce the function $\mathscr{F}$ and the corrections due to $(V_P - V_B)$ are obtained iteratively in a Born-series manner as an approximation to the exact function ψ. The asymptotic value of $\mathscr{F}$ is much closer to that of ψ than for method $\mathscr{S}_2$, but the rate of convergence of the Green's function iterations is not as fast as that of method $\mathscr{S}_2$, using potential V_B (or V_S) for generating the Sturmian functions. Contrary to what is the case for a general integral kernel $\mathscr{O}$, the Green's function iterative method can only be used for the case when $\mathscr{O} = \mathscr{G}_0 V$, while the method based on Eqs. (11.42)–(11.46) is more general.

An examination of the uniformity of the convergence of the iterations shows that the convergence at a distance smaller than the range of the Sturmian potential is significantly better than at larger distances. This phenomenon is illustrated in Fig. 11.16, and is due to the gradual loss of independence between the Sturmian functions at large distances. Had the potential V_S been used for generating the Sturmian functions for Fig. 11.16, rather than Sturmian potential V_{WS}, then after the 14th iteration the error of the wave function would have become large even for $r > 7$ fm, and asymptotically the error would have been several orders of magnitude larger than the error shown in Fig. 11.16. Both Figs. 11.14 and 11.16 attest to the importance of using a basis of Sturmian functions generated with an auxiliary potential whose range is larger than (or at least equal to) the range of the operator $\mathscr{O}$ in Eq. (11.27).

For very long-ranged potentials, such as a Coulomb potential (that decreases as $1/r$), or a dipole-dipole interaction potential (that decrease like $1/r^3$), the calculation of Sturmian functions can be carried out by invoking the Phase-Amplitude method

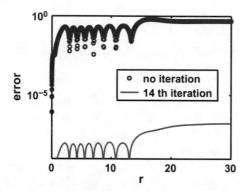

Fig. 11.16 Absolute value of the error of the wave function as a function of radial distance r. The result labeled as "no iteration" illustrates the result for $\mathscr{F}^{(2)}$, obtained by solving Eq. (11.49), with $N = 31$. The other line is obtained after the 14th iteration. Method $\mathscr{S}_2$ was used for these results, using positive energy Sturmians for potential V_{WS}. The "exact" scattering function ψ, which provides a measure of the error of the iteration results, is obtained with potential V_P, and $k = 0.5\,\mathrm{fm}^{-1}$, using the spectral integral equation method S-IEM

for the long-distance part of the wave function [26], where it is carried out with the spectral expansion method. Improvements for this method, however, are still under development.

11.7 Conclusions

Although the negative energy Sturmians have been used during many years to obtain a separable representation of potentials or of two-body T-matrices, the use of positive energy Sturmians for the solution of atomic physics problems has been revived only recently through the work of the Argentinian team we mentioned before [14–19]. This group successfully extended the application to problems with more than one variable, and with more particles present.

The review of the properties of Sturmian functions, and of the solution of integral equations with a general integration kernel, has confirmed that the convergence of the Sturmian expansion is very slow [21] compared to spectral Chebyshev expansions, and hence iterative corrections are needed to make up for the truncation error. We have presented two correction schemes, and have shown that approximately 20 iterations were needed to provide accuracies on the order of $1 : 10^{-5}$.

References

1. D. Mitnik, F.D. Colavecchia, G. Gasaneo, J.M. Randazzo, Comput. Phys. Commun. **182**, 1145–1155 (2011)
2. C.M. Bender, S.A. Orszag, *Advanced Mathematical Methods for Scientists and Engineers* (McGraw Hill Inc, New York, 1978)
3. J.C. Guillou, J. Zinn-Justin, *Large-Order Behaviour of Perturbation Theory* (Elsevier B.V., Amsterdam, 2016). ISBN 978-00444-88597-5; G.V. Dunne, *Perturbative - Non-perturbative Connection in Quantum Mechanics and Field Theory* (2002), arXiv:hep-th/0207046
4. B.D. Shizgal, *Spectral Methods in Chemistry and Physics. Applications to Kinetic Theory and Quantum Mechanics* (Springer, Dordrecht, 2015)
5. S. Kallush, R. Kosloff, F. Masnou-Seeuws, Phys. Rev. A **75**, 043404 (2007); O. Dulieu, R. Kosloff, F. Masnou-Seeuws, G. Pichler, J. Chem. Phys. **107**, 10633–10642 (1997)
6. S. Weinberg, Phys. Rev. **131**, 440 (1963); ibid., **133B**, 232 (1964); S. Weinberg, M. Scaldron, ibid., **133B**, 1589 (1964); S. Weinberg, *Lectures on Particles and Field Theory, Brandeis Summer Institute in Theoretical Physics*, vol. 2 (Prentice-Hall, Englewood Cliffs, 1964)
7. M. Rotenberg, Ann. Phys. (N.Y.) **19**, 262 (1962); M. Rotenberg, Theory and applications of Sturmian functions, in *Advances in Atomic and Molecular Physics*, vol. 6, ed. by D.R. Bates, I. Esterman (Academic, New York, 1970), pp. 233–268
8. S. Klarsfeld, A. Maquet, Phys. Lett. A **78**, 40 (1980)
9. E. Karule, A. Gailitis, J. Phys. B: At. Mol. Opt. Phys. **43**, 065601 (2010)
10. H. Rottke, B. Wolff-Rottke, D. Feldmann, K.H. Welge, M. Dorr, R.M. Potvliege, R. Shakeshaft, Phys. Rev. A **49**, 4837–4851 (1994); S.Y. Ovchinnikov, J.H. Macek, Phys. Rev. A **55**, 3605 (1997); S.Yu. Ovchinnikov, G.N. Ogurtsov, J.H. Macek, Yu.S. Gordeev, Phys. Rep. **389**, 119 (2004)
11. K. Amos, L. Canton, G. Pisent, J.P. Svenne, D. van der Knijff, Nucl. Phys. A **728**, 65 (2003); P. Fraser et al., Phys. Rev. Lett. **101**, 242501 (2008); U. Weiss, Nucl. Phys. A **156**, 53 (1970)
12. L.N. Trefethen, *Spectral Methods in MATLAB* (SIAM, Philadelphia, 2000); J.P. Boyd, *Chebyshev and Fourier Spectral Methods*, 2nd revised edn. (Dover Publications, Mineola, 2001); B. Fornberg, *A practical Guide to Pseudospectral Methods*. Cambridge Monographs on Applied and Computational Mathematics (Cambridge University Press, Cambridge, 1998)
13. R.A. Gonzales, J. Eisert, I. Koltracht, M. Neumann, G. Rawitscher, J. Comput. Phys. **134**, 134–149 (1997); R.A. Gonzales, S.-Y. Kang, I. Koltracht, G. Rawitscher, J. Comput. Phys. **153**, 160–202 (1999); G. Rawitscher, I. Koltracht, Comput. Sci. Eng. **7**, 58 (2005); G. Rawitscher, Applications of a numerical spectral expansion method to problems in physics: a retrospective, in *Operator Theory, Advances and Applications*, vol. 203, ed. by T. Hempfling (Birkäuser Verlag, Basel, 2009), pp. 409–426; A. Deloff, Ann. Phys. (NY) **322**, 1373–1419 (2007)
14. G. Gasaneo, L.U. Ancarani, J. Phys. **45**, 045304 (2012)
15. A.L. Frapiccini, J.M. Randazzo, G. Gasaneo, F.D. Covalecchia, Phys. Rev. A **82**, 042503 (2010)
16. M. Ambrosio, D.M. Mitnik, G. Gasaneo, J.M. Randzzo, A.S. Kadyrov, D.V. Fursa, I. Bray, Phys. Rev. A **92**, 052518 (2015)
17. A.L. Frapiccini, J.M. Randazzo, F.D. Colavecchia, J. Phys. B **43**, 101001 (2010)
18. D.M. Mitnik, G. Gasaneo, L.U. Ancarani, M.J. Ambrosio, J. Phys. Conf. Ser. **488**, 012049 (2014)
19. G. Gasaneo, D.M. Mitnik, A.L. Frapiccini, F.D. Colavecchia, J.M. Randazzo, J. Phys. Chem. A **113**, 14573 (2009)
20. G. Rawitscher, Phys. Rev. C **25**, 2196 (1982)
21. L.R. Dodd, Phys. Rev. A **9**, 637 (1974); I.H. Sloan, Phys. Rev. A **7**, 1016 (1973)
22. D.M. Mitnik, F.D. Colavecchia, G. Gasaneo, J.M. Randazzo, Comput. Phys. Commun. **182**, 1145 (2011)
23. G. Rawitscher, Phys. Rev. E **85**, 026701 (2012)
24. F. Perey, B. Buck, Nucl. Phys. **32**, 353 (1962)
25. J. Power, G. Rawitscher, Phys. Rev. E **86**, 066707 (212)
26. G. Rawitscher, Comput. Phys. Commun. **191**, 33–42 (2015)

Chapter 12
Solutions of a Third-Order Differential Equation

Abstract In this chapter we show that spectral methods can be applied to a third-order differential equation. Such an equation can be solved by a Collocation method with Chebyshev polynomials and without the application of Green's functions.

12.1 The Objective and Motivation

In Chap. 6 we described the solution of a second order differential equation in terms of an equivalent integral equation, using a Green's function method that also embodied the boundary condition of the solution. The procedure was based on a spectral collocation method using Chebyshev polynomials as the expansion basis. In Chap. 7 we described a solution based on a Galerkin method using Lagrange functions for the expansion basis. In the present chapter we describe the solution of a third-order differential equation using a collocation method, and also considering Chebyshev polynomials for the expansion basis, but not using Green's functions, nor equivalent integral equations. The purpose is to provide a contrast to the Galerkin method, and to the integral equation method, the latter of which is not applicable in this case. The third order linear equation being solved is for the square of the amplitude of the oscillatory solution of a second order differential equation. This chapter is mainly based on Ref. [1].

12.2 Introduction

In Chap. 8 we presented a phase-amplitude description of a wave function $\psi(r)$, which for the case when the potential V is smaller than the energy E, is an oscillatory function of distance r. In this case $\psi(r) = y(r) \sin[\phi(r)]$, where y is the amplitude and $\phi(r)$ is the phase. Milne's [2] equation for $y(r)$ is non-linear, as given

© Springer Nature Switzerland AG 2018
G. Rawitscher et al., *An Introductory Guide to Computational Methods for the Solution of Physics Problems*,
https://doi.org/10.1007/978-3-319-42703-4_12

by $d^2y/dr^2 + k^2y = V\,y + k^2/y^3$. A solution was obtained by an iterative method, as described in Chap. 8. However the iterations do not converge for r located near a turning point, and hence an alternate method became desirable. Such an alternate method is explored in this chapter for the solution of a linear equation for the square of the amplitude $\eta(r) = y^2(r)$. This linear third-order differential equation for the square of the amplitude $\eta(r)$ can be derived from Milne's non-linear equation. The purpose of the present chapter is to explore the properties of the solution of this equation. We find that in the vicinity of turning points this equation leads to unstable solutions.

12.3   The Third-Order Linear Equation

By starting from Eq. (8.3) for either when $E \leq V$ or when $E \geq V$, and by defining

$$\eta(r) = y^2(r), \tag{12.1}$$

one finds that η obeys the linear third-order differential equation.

$$\eta''' + 4(E - V)\eta' - 2V'\eta = 0, \tag{12.2}$$

where the "prime" means a derivative with respect to r. The proof is as follows [3]:
   (a) One defines $y = \pm\eta^{1/2}$ and calculates both y' and y'' in terms of the derivatives of η. After equating the result to either Eq. (8.5) or (8.9) for the cases $E > V$ or $E < V$, respectively, and after multiplying both sides by $\eta^{(3/2)}$, one obtains the result

$$-\frac{1}{4}(\eta')^2 + \frac{1}{2}\eta\eta'' = \pm k^2 + (V - E)\eta^2. \tag{12.3}$$

   (b) This equation is still non-linear. The linearization occurs when taking the derivative with respect to r of the terms in Eq. (12.3). After some cancellations, Eq. (12.2) is produced.
   Once the function $y^2(r)$ is obtained the phase can be calculated according to

$$\phi(r) = k \int_0^r \frac{1}{y^2(r')}dr' + \phi(0), \tag{12.4}$$

where k is the wave number, given by $k = E^{1/2}$ when $E > 0$. From these ingredients the Ph-A wave function ψ_{P-A} can be obtained according to

$$\psi_{P-A} = \psi_{P-A} = \begin{cases} y\sinh(\phi) & \text{for } E > V \\ y\exp(-\phi) & \text{for } E < V \end{cases}. \tag{12.5}$$

12.4 Iterative Solution of Eq. (12.2)

Since the third-order derivative of the function $y^2 = \eta(r)$ is small, and goes to zero for large distances (because y^2 approaches a constant), one can use this fact to set up an iterative solution of Eq. (12.2). The procedure consists in rewriting Eq. (12.2) in the form

$$(E - V)\eta' - \frac{1}{2}V'\eta = \chi(r), \tag{12.6}$$

with

$$\chi(r) = -\frac{1}{4}\eta''', \tag{12.7}$$

and treat the right hand side χ of Eq. (12.6) as a perturbation.

The first approximation consists in setting $\chi = 0$. In this case a solution of $(E - V)\eta' - V'\eta/2 = 0$ is given by the WKB expression [4, 5] η_{WKB} for the amplitude.

$$\eta_{WKB}(r) = \frac{\sqrt{E}}{\sqrt{E - V(r)}}. \tag{12.8}$$

The constant $\sqrt{E}$ in the numerator is chosen such that asymptotically $\eta_0(r \to \infty) \to 1$. Here it is assumed that $E - V > 0$, and $E > 0$, but the procedure can be adapted to the case that $E - V < 0$. However the iteration will not converge in the vicinity of turning points, where $E - V \simeq 0$.

The next step consist in solving the first order inhomogeneous equation

$$(E - V)\eta'_{n+1} - \frac{1}{2}V'\eta_{n+1} = \chi_n(r), \quad n = 1, 2, 3, \ldots \tag{12.9}$$

with the iteration index n set to 1, and $\chi_1(r) = -\frac{1}{4}\eta'''_{WKB}$. The third order derivative of η_{WKB} is calculated by expanding η_{WKB} in terms of Chebyshev polynomials $\eta_{WKB}(r) = \sum_{n=1}^{N+1} c_n^{(WKB)} T_{n-1}(x)$, and obtaining

$$\eta'''_{WKB}(r) = \sum_{n=1}^{N+1} c_n^{(WKB)} T'''_{n-1}(x), \tag{12.10}$$

where a "prime" means derivative with respect to r. Numerical examples of the use of Eq. (12.10) are presented in Chap. 5. For $n > 1$, the procedure is repeated iteratively.

An analytic solution of Eq. (12.9) can be achieved by introducing an integration factor $\Omega(r)$ such that

$$\{\Omega(E - V)\eta\}' = \Omega \left[(E - V)\eta' - \frac{1}{2}V'\eta) \right], \tag{12.11}$$

with the result (after some algebra)

$$\Omega(r) = \frac{c}{\sqrt{E-V}}, \tag{12.12}$$

where c is an arbitrary constant that cancels in the end. Hence, Eq. (12.9) becomes $\left[\Omega(E-V)\eta_{n+1}\right]' = \Omega\chi_n$, whose solution is given by the analytic expression

$$\eta_{n+1}(r) = \frac{1}{\sqrt{E-V(r)}} \int_a^r \frac{1}{\sqrt{E-V(r')}} \chi_n(r')dr', \ n = 1, 2, \ldots, n_{max}. \tag{12.13}$$

The normalization of η consists in obtaining a renormalization constant c_1 such that $\eta_{n+1} = c_1\eta_{n+1}$, with $\eta_{n+1}(r_{max}) = \eta_{WKB}(r_{max})$, where r_{max} is the largest distance beyond which the numerical calculation is stopped, and a is the starting value of r, denoted as r_{start} below. Please note that only integrals, together with the use of Eq. (12.10), are needed to implement the solution.

12.4.1 Coulomb Example

The rounded Coulomb potential is again used in this example for comparison with the solution of Eq. (12.38). This potential has been defined in Eqs. (8.34)–(8.36). The numerical results for $y(r)$ after 7 iterations are given in Fig. 12.1. The input parameters are $\bar{Z} = 4$, a rounding parameter of $t = 2$, a starting value of $r_{start} = 5$ and $r_{max} = 150$, and a number $N = 61$ of Chebyshev expansion polynomials, and an energy $E = 1$. Near the turning point for this potential, which for this energy occurs at $r_{turn} \simeq 3.3$, the WKB approximation for y and the iterated value differ substantially.

It is clear that the iterations converge, and the resulting value of y^2, illustrated in Fig. 12.1, does not present the zeros shown in the case of positive energies. However the iterative solution of Eq. (12.2) does not converge near the turning points, because

Fig. 12.1 The result for the amplitude $y = \sqrt{\xi}$ obtained from Eq. (12.13) after the 7th iteration for the rounded Coulomb potential described in the text. The calculation is started at $r = 5$

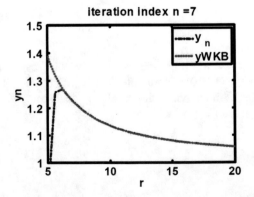

the term $1/\sqrt{E - V(r)}$ in Eq. (12.13) becomes too large. We describe a non-iterative method next.

12.5   A Chebyshev Collocation Solution

The purpose of this approach is to avoid the use of Green's function in the S-IEM solution described in Chap. 6 and to avoid using an iterative solution described in Sect. 12.4 (because it does not converge near the turning points). The method, denoted by "CCM", consists in doing a Chebyshev expansion of the square of the amplitude and imposing the boundary condition at large distances by matching it there to the WKB approximation. The procedure is described below, and it differs markedly from the procedure described in Ref. [6], given that the latter is stable in the vicinity of turning points. In Fig. 12.2 we show the rate of convergence of Eq. (12.13) as expressed in terms of the difference between successive iteration values n.

The function $\eta(r) = y^2(r)$ is expanded in a set of Chebyshev polynomials $T_v(x)$, $v = 0, 1, 2, \ldots, N$, with $-1 \le x \le 1$, and $b_1 \le r \le b_2$, where r is mapped into x by a linear transformation. The expansion is given by

$$\eta(r) = \sum_{n=1}^{N+1} a_n T_{n-1}(x),\tag{12.14}$$

and Eq. (12.2) takes the form

$$\hat{O}\eta(r) = \sum_{n=1}^{N+1} a_n \hat{O} T_{n-1}(x),\tag{12.15}$$

with the operator $\hat{O}$ given by

$$\hat{O} = \frac{d^3}{dr^3} + 4(E - V)\frac{d}{dr} - 2\frac{dV}{dr}.\tag{12.16}$$

Fig. 12.2 The rate of convergence of Eq. (12.13) as expressed in terms of the difference between successive iteration values n of y. The latter are obtained form the normalized values of $\bar{\xi}_n$, and $y^{(n)} = \sqrt{\xi^{(n)}}$

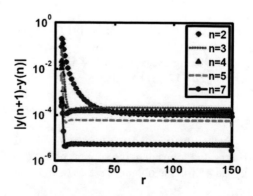

Matching to the WKB solution in the vicinity of b_2, at points r_1 and r_2 (which correspond to points x_1 and x_2, respectively) is accomplished by the requirement

$$\eta(r_i) \equiv \eta_{WKB}(r_i) = \sum_{n=1}^{N+1} a_n T_{n-1}(x_i), \, i = 1, 2, \tag{12.17}$$

This condition can be used to express a_1 and a_2 in terms of $a_3, a_4, \ldots, a_{N+1}$. In the matrix form this condition can be expressed as

$$M_1 A_1 + M_3 A_3 = \eta_{WKB}, \tag{12.18}$$

where

$$M_1 = \begin{bmatrix} T_0(x_1) \, T_1(x_1) \\ T_0(x_2) \, T_1(x_2) \end{bmatrix}, \tag{12.19}$$

$$M_3 = \begin{bmatrix} T_3(x_1) \, T_4(x_1) \, \ldots \, T_N(x_1) \\ T_3(x_2) \, T_4(x_2) \, \ldots \, T_N(x_2) \end{bmatrix}. \tag{12.20}$$

The column vectors A_1 and A_3 are given by

$$A_1 = (a_1, a_2)^T, \quad A_3 = (a_3, a_4, \ldots, a_{N+1})^T, \tag{12.21}$$

where the superscript T means "Transposed", and the column vector η_{WKB} is given by

$$\eta_{WKB} = (\eta(r_1), \eta(r_2))^T. \tag{12.22}$$

The Eq. (12.15) can likewise be expressed in matrix form

$$O_1 A_1 + O_3 A_3 = 0, \tag{12.23}$$

where

$$O_1(x) = \begin{bmatrix} \hat{O}T_0(x) \, \hat{O}T_1(x) \end{bmatrix}, \tag{12.24}$$

$$O_3(x) = \begin{bmatrix} \hat{O}T_3(x) \, \hat{O}T(x) \, \ldots \, \hat{O}T_N(x) \end{bmatrix}. \tag{12.25}$$

By making use of Eq. (12.18), the vector A_1 can be expressed in terms of A_3. As a result Eq. (12.23) becomes

$$\left[-O_1(x)M_1^{-1}M_3 + O_3(x) \right] A_3 = -O_1(x)M_1^{-1}\eta_{WKB}. \tag{12.26}$$

In the equation above, x has a general value in the domain $[-1, 1]$. If that equation is repeated for each of the support points $\xi_i, i = 1, 2, \ldots, N + 1$, as given in Chap. 5 by Eq. (5.6), one obtains a set of $N - 1$ linear equations for $a_3, a_4, \ldots, a_{N+1}$, which when solved produces the vector A_3. The vector A_1 can be obtained in terms of A_3 by means of equation

$$A_1 = M_1^{-1}(\eta_{WKB} - M_3 A_3),\tag{12.27}$$

and the function $\eta(r)$ can be obtained in terms of Eq. (12.22). The Eqs. (12.26) and (12.27) express the final result. Both equations now contain the asymptotic boundary condition given by η_{WKB}, and do not involve the use of Green's functions. In the subsection below we present numerical applications to the case of a repulsive rounded Coulomb potential, a potential with an attractive valley and a repulsive barrier. For these examples the functions $\eta(r)$ do not have zeros, they change slowly with r, but become unstable in the vicinity of the turning points.

12.5.1   A Numerical Example for an Attractive Coulomb Potential

The application to the rounded Coulomb potential is described below. This potential is defined in Chaps. 5–7, where the Coulomb potential is replaced by $Ze^2/\mathscr{R}(r)$, with $\mathscr{R}(r)$ given by

$$\mathscr{R}(r) = r/[1 - \exp(-r/t)].\tag{12.28}$$

This "rounding" procedure eliminates the singularity at the origin but does not alter the point Coulomb potential at distances much larger than the rounding parameter t. This potential is introduced into the renormalized Schrödinger equation (8.5), and the renormalized potential is given by

$$V(r) = \frac{Ze^2}{\mathscr{R}(r)}\frac{2m}{\hbar^2} = \frac{\bar{Z}}{\mathscr{R}(r)}.\tag{12.29}$$

Here, $\bar{Z}$ has units of inverse length and is related to the Coulomb parameter η according to

$$\bar{Z} = 2\eta k.\tag{12.30}$$

For repulsive (attractive) potentials, $\bar{Z}$ is positive (negative), and in our example $\bar{Z} < 0$.

A result with $E = 1$, an attractive potential with $\bar{Z} = -4$, a rounding parameter $t = 2$ and a radial interval $5.5 \leq r \leq 40$, in which 30 Chebyshev expansion functions are used, is illustrated in Fig. 12.3.

By matching the result of Fig. 12.3 at $r_1 = 5.1$ and $r_2 = 5.6$ to an additional calculation from $r = 2$ to $r = 6$, and using between 7 and 13 Chebyshev support points, one obtains the result shown in Fig. 12.4.

What is noteworthy about Fig. 12.4 is that the result for y is a smooth function valid in the vicinity of the turning point at $r = 3.3$, and all of the curves reach a

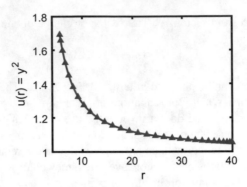

Fig. 12.3 The radial domain is [5.5, 40], and the number of expansion Chebyshev polynomials is $N + 1 = 41$

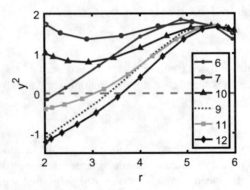

Fig. 12.4 The results for y^2 displayed in this graph are obtained by starting the solution at points near 5.0 and 5.6, where the values from Fig. 12.3 are used as input numbers. The number of Chebyshev polynomials is incrementally changed from 7 to 13. The purpose of this figure is to show that the solutions are not very stable in this region encompassing the turning point that occurs at $r \simeq 3.3$

maximum and then decay. The fact that the solutions strongly depend on the number of expansion polynomials used shows that the solution is unstable in the vicinity of a turning point. The results labeled with "7" or "10" are compatible with Fig. 12.1 near the maximum of y. In order to further justify this statement Fig. 12.5 illustrates the comparison of the amplitude y with the wave function ψ, which shows that the maximum of the wave function agrees with the value of y (as should be the case). Here ψ is the wave function obtained from solving the Schrödinger equation by means of the S-IEM method. By contrast, the result shown in Fig. 12.3 is non-oscillatory and very stable.

In a section below, additional arguments are presented to show that in the vicinity of turning points Eq. (12.2) is ill-conditioned.

Table 12.1 lists the accuracy for the function $\eta(r)$ at various values of the number $N + 1$ of Chebyshev polynomials used for the expansion of $\eta(r)$ in the radial

Fig. 12.5 Comparison of the peak of the wave function with the value of y for two values of the number of Chebyshev polynomials used in the radial interval [2, 6]

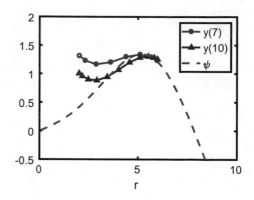

Table 12.1 Accuracy of $u(r)$ for the attractive Coulomb potential

N	Cond #	Accuracy
100	2×10^5	$1 : 10^{11}$
40	3×10^3	$1 : 10^7$
20	5×10^2	$1 : 10^4$

interval [$5.5 \le r \le 200$], and also the value of the condition number of the matrix $\left[-O_1(x)M_1^{-1}M_3 + O_3(x) \right]$ used in Eq. (12.26), for the case of the attractive rounded Coulomb potential described in the present section. This accuracy estimate is based on the size of the absolute value for the last three expansion coefficients a_n in Eq. (12.14).

12.5.2   Numerical Example for an Atomic Physics-Type Potential

The potential we use here has an attractive valley near the origin, a repulsive barrier of maximum height of $\simeq 6$ in units of inverse length squared at $r \simeq 4$, and continues to decrease exponentially as $r \rightarrow \infty$. The example illustrated in Fig. 12.6 mimics the potentials describing atom-atom interactions, however the numbers (in units of inverse length squared) are not realistic. This potential is given by Eq. (12.31)

$$V(r) = -6 \, y(y - 2); \quad y = \exp(-0.3r + 1.2). \qquad (12.31)$$

In order to avoid the occurrence of turning points, the energy in this numerical example is above the height of the barrier, $E = (2.8)^2$.

A comparison between the value of the amplitude y and the Schrödinger wave function, both for $L = 0$ and an energy of $(k = 2.8)^2 \simeq 7.9$ is illustrated in Fig. 12.7. The good agreement between the tips of $|\psi|$ and the values of y again confirms the validity of the *CCM* method of solution of Eq. (12.2). In the valley of the potential

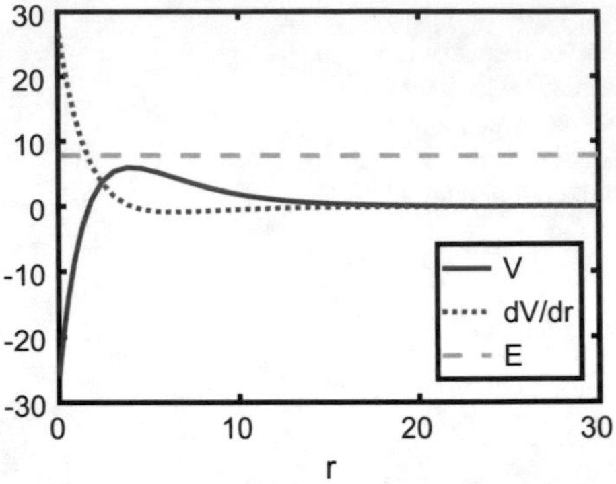

Fig. 12.6 This potential is
defined by Eq. (12.31)

Fig. 12.7 The quantity y is
obtained from the solution of
Eq. (12.2) for the Morse-type
potential described by
Eq. (12.31), for an energy of
$(2.8)^2$, a radial domain
$[0, 30]$, and with 41
Chebyshev support points

Fig. 12.8 The phase as a
function of distance r, for
the atomic-physics-type
potential. The calculation is
based on Eq. (12.4), and the
parameters are the same as
for Fig. 12.6

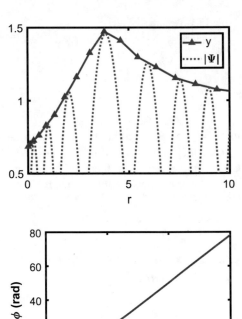

near the origin, the distance between the maxima of $|\psi|$ is small compared to the
distance in the barrier region. The local wave number is large in the valley of the
potential, and near the top of the barrier it is smaller, and for larger distances it
becomes larger again. This can be confirmed by the evaluation of the phase given by
Eq. (12.4). The result for the phase with the same parameters as used for Fig. 12.7 is
displayed in Fig. 12.8.

This figure shows that in the radial region near the origin where the potential is attractive, the slope of ϕ is large, which means that the local wave number is small. In the barrier region, for $3 < r < 7$, the curve bends sightly towards the horizontal axis, which indicates that the wave number increases with r more slowly, while for $r > 7$ the curve continues with a larger slope. These features are in accordance with Fig. 12.8.

In conclusion, the CCM for $E > V$ proved to be reliable and in agreement with the respective oscillatory wave functions ψ. The CCM is apparently a new method.

12.6 Solution of Eq. (12.2) by a Finite Difference Method

We will now show that the Finite Difference Method provides an unstable solution to Eq. (12.2) in the vicinity of a turning point. By using the finite difference equations described in Chap. 2, based on Taylor's expansion to 5th order, one finds that

$$\eta_{n+1} - \eta_{n-1} = 2h\, \eta_n' + \frac{1}{3}h^3\, \eta_n''' + O(h^5). \tag{12.32}$$

The support points in the radial interval are equi-spaced, with the distance between two neighboring points given by h, each point is given by $r_n = n\,h$, $n = 1, 2, \ldots$, and $\eta_n = \eta(r_n)$. Similarly $V_n = V(r_n)$. By rewriting Eq. (12.2) in the form

$$\eta_n''' = A_n\, y_n' + 2\, V_n' \eta_n, \tag{12.33}$$

with and inserting this expression into Eq. (12.32), one finds

$$\eta_{n+1} - \eta_{n-1} = \eta_n' \left(2h + \frac{1}{3}h^3 A_n\right) + \frac{2}{3}h^3 V_n' \eta_n + \mathcal{O}(h^5). \tag{12.34}$$

Finally, by making use in Eq. (12.34) of

$$\eta_n' = \frac{\eta_{n+1} - \eta_{n-1}}{2h} + \mathcal{O}(h^2), \tag{12.35}$$

one obtains

$$\eta_{n-1} = \eta_{n+1} + 4h\frac{V_n'}{A_n}\eta_n + \mathcal{O}(h^3). \tag{12.36}$$

Since A_n approaches $\simeq 0$ in the vicinity of a turning point, the finite difference method becomes unstable.

In conclusion, the advantage of the third order linear differential equation is that in the radial region devoid of turning points an efficient method of solution is feasible with the spectral Chebyshev collocation expansion method described in this section. That method is not iterative, has good accuracy, permits the calculation of wave functions out to large distances and of course it can be applied to other cases. However, in the vicinity of turning points this method becomes unstable.

12.6.1 Second-Order Linear Equation

One can also proceed by transforming Eq. (12.2) into a second order equation. The steps are as follows. One defines the variable $v(r)$ as

$$v(r) = d\eta/dr \tag{12.37}$$

and one arrives at the second order linear differential-integral equation

$$d^2v/dr^2 + 4(E - V)v = 2(dV/dr)\left(\int_0^r v(r')dr'\right). \tag{12.38}$$

Here $k^2 = E$ is the energy and $V(r)$ the potential in the Schrödinger equation. Once the function $v(r)$ is obtained, then the function $y^2(r)$ is also obtained according to

$$\eta(r) \equiv y^2(r) = \int_0^r v(r')dr', \tag{12.39}$$

and the phase can be calculated according to Eq. (12.4).

At large distances the right hand side of Eq. (12.38) and the potential V in the left hand side both become negligible, and Eq. (12.38) reduces to

$$d^2v/dr^2 + 4Ev \simeq 0. \tag{12.40}$$

For positive energies the solution of Eq. (12.40) is an oscillatory function of r, and for negative energies it is an exponential function. The corresponding function η, being an integral of the former, also becomes oscillatory, and hence is not useful since it is supposed to be a smooth nearly monotonic function of r. This result is illustrated by means of a Coulomb potential example given in Fig. 12.9.

A method of solving Eq. (12.38) consists in using a Green's function spectral Chebyshev method described in Chap. 6. A detailed description of the solution and of the results is given in Ref. [7]. The Green's function was the "inverse" of the operator $d^2/dr^2 + 4E$, and the boundary conditions at asymptotic values of r are given by the WKB expression for η. It was found for a numerical rounded Coulomb potential example with $k = 1$ (no turning points) that $\eta(r)$ was non-monotonic as illustrated in Fig. 12.9 [7], and hence is not very useful.

However a feature in favor of Eq. (12.38) is that the solution is stable in the vicinity of turning points.

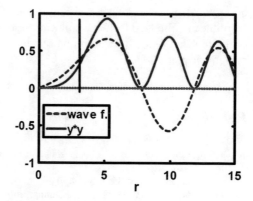

Fig. 12.9 The wave function ψ, and the function y^2, for the case of a repulsive rounded Coulomb potential, with $\bar{Z} = 4$, and a positive energy of $E = 1$. The position of the turning point is marked by the vertical line

12.7 Summary and Conclusion

The main results of the present chapter are:

1. The solution of Eq. (12.2) by the Chebyshev Collocation Method (*CCM*) removes the need for performing an iterative solution of the original Milne's non-linear equation [2] for the amplitude y, provided it is carried out in a radial region devoid of turning points. By means of the phase-amplitude description, that condition enables one to obtain an accurate and efficient solution of a one-dimensional Schrödinger equation out to large distances.

2. The solution of Eq. (12.2) becomes unstable in the vicinity of turning points. This is shown both via the *CCM* and a finite difference method.

3. The *CCM* is general, and can be applied to other types of differential equations.

In summary, the advantage of the third order linear differential equation (12.2) is that in the radial region devoid of turning points the solution by means of the *CCM* provides an efficient and simple algorithm as we have described in this book. It is a new method, is not iterative, has good accuracy, and permits the calculation of wave functions out to large distances. The development we have presented opens a practical way for applying the Ph-A description to many physics and engineering situations.

12.8 Project 12.2

Repeat the calculations described in Sect. 12.5 for the long range region [6, 40], but evaluate the wave function ψ from the solution of the Schrödinger equation given by Eq. (12.2) in the short range region [0, 6] by a finite difference method, such as Runge–Kutta or Numerov. Then by a method explained in Sect. 8.3.5, obtain the phase and amplitude of ψ in the vicinity of $r \simeq 6$. A normalization of ψ may be required, so as to match the amplitude obtained from ψ near $r = 6$ with the amplitude y^2 obtained from the collocation solution. If possible, use for the potential V the rounded repulsive Coulomb potential given in the numerical example of the present chapter.

References

1. G. Rawitscher, Solution to a linear equation for the phase amplitude description of a wave-function, submitted to J. Phys. B
2. W.E. Milne, Phys. Rev. **35**, 863 (1930)
3. The authors thank Professor Gerald Dunne, from the Department of Physics at UCONN, for calling their attention to the existence of this equation
4. H. Jeffreys, B.S. Jeffreys, *Methods of Mathematical Physics* (Cambridge University Press, Cambridge, 1966)
5. H.A. Kramers, Z. Phys. **39**, 828 (1926)
6. G. Rawitscher, C. Merow, M. Nguyen, I. Simbotin, Am. J. Phys. **70**(9), 935–944 (2002)
7. G. Rawitscher, Few-body Systems **58**, 5 (2017)

Chapter 13
Conclusions

Abstract We discuss here the main topics developed in this monograph. Through our analysis, examples and calculations, we show that spectral methods present various advantages in relation to the methods based on finite differences in terms of accuracy, convergence, time of computation and stability.

13.1 Overview

It is the objective of this work to provide a monograph about numerical methods for solving differential or integral equations by using finite difference schemes, finite element techniques and the most efficient spectral methods. The monograph places special emphasis on the latter. Our work is to be useful to undergraduate and graduate students. In terms of undergraduate courses, the first two chapters are focused on reviewing or learning basic types of errors in numerical calculation, convergence and stability of solutions of iterative equations and orders of finite difference methods, all of which the monograph illustrates with simple classical examples. For graduate courses, the most advanced chapters are useful for students who want learn new methods including the S-IEM, FE-DVR or CCM methods described in the text. Undergraduate students who are not familiar with spectral methods can start with the description of the Galerkin and Collocation Methods and can go deeper in the concepts studying the theorems on spectral methods, the description of spectral methods in different equations and their applications to interesting physics problems.

Throughout the text we have shown that spectral methods offer significant advantages over other methods, and have indicated how these methods can be applied to obtain these types of improvements. We draw special attention to the following specific conclusions and techniques that we have developed in this monograph which provide marked advantages for scientists working in the field of spectral methods:

- Round-off and truncation errors cannot be avoided in numerical calculations and spectral methods lead to very good accuracy, are fast and are not so complex as thought.

© Springer Nature Switzerland AG 2018
G. Rawitscher et al., *An Introductory Guide to Computational Methods for the Solution of Physics Problems*,
https://doi.org/10.1007/978-3-319-42703-4_13

- Spectral methods are based on expansions in a set of basis functions They are very efficient, especially when we consider Chebyshev polynomials in the calculations.
- Two main forms of the solutions of the linear equation $Lu = f$ can be implemented in terms of expansions into a set of basis functions: Galerkin and Collocation methods, which also provide solutions with higher accuracy than when finite difference methods are used.
- Cases of non-linear equations cannot be solved by Galerkin and Collocation methods presented, but they can be associated with iterative procedures to produce solutions.
- Theorems guarantee that in expansions of a function $f(x)$ into a truncated set of $N + 1$ Chebyshev polynomials, the interpolation error can be determined and show that the Chebyshev mesh points give rise to the best interpolation.
- A numerical solution of Lippmann–Schwinger (L–S) equation equivalent to a second order differential equation presents many advantages. By working with that integral equation, one can expand the solution in terms of Chebyshev polynomials (spectral integral equation method or S-IEM) and a good accuracy can be obtained.
- A spectral finite element method (FE-DVR) can be constructed to obtain the solution of second order linear differential equations by means of expansions into a set of Lagrange polynomials (discrete variable representation or DVR) when the radial domain is subdivided into contiguous partitions. Its accuracy is good, but is lower than that obtained by the S-IEM method.
- One can solve the Schrödinger equation by applying the Phase-Amplitude Method (Ph-A), which provides very reliable results. The spectral method for iteratively solving the non-linear equation of the amplitude function in Ph-A is efficient due to the small number of mesh points required, and the much smaller rate of accumulation of errors when compared with the rate when one directly obtains the wave function.
- One can describe the propagation of waves on inhomogeneous strings by applying either the Galerkin-Fourier expansion method or the spectral Greens function Collocation method to obtain the solution of a Sturm–Liouville eigenvalue equation. In addition, our study shows that the latter may be preferable for most applications, although this method is more complex than than the former.
- One can calculate eigenvalues of a second order differential equation by the spectral S-IEM method and the iterations converge very quickly to high accuracy.
- Sturmian functions can also be adopted in order to solve integral equations with a general integration kernel. Our study confirms that the convergence of the Sturmian expansions is slow, but their accuracy can be improved by using spectral methods based on Chebyshev expansions In addition, iterative corrections are needed to provide high accuracies.
- A third-order differential equation such as the Milnes non-linear equation can be solved without requiring iterations from Collocation methods with Chebyshev expansion (Chebyshev Collocation Method or CCM). The results are very accurate.

13.2 Perspectives and Final Consideration

Spectral methods constitute a wide and growing field that in the future can offer many novel and useful solutions to problems in Physics. Our experience of decades of research in the field of numerical calculations has enabled us to present in this monograph what we consider to be some of the most important and useful aspects of spectral methods. In addition to those methods that we described here, there are a number of promising areas for students and scientists in the development of the field. Some of the important topics that were left out of this monograph include:

- The extension of the spectral methods to many dimensions;
- The solution of physics problems other than those described by the Schrödinger wave equation;
- A more extensive comparison of the finite element method that use expansion functions in each element with splines functions, or Gaussian function.
- The treatment of Stiff equations and other differential or integral equations, among other cases.

In summary, our main purpose in this project was to demonstrate the large improvement over finite difference methods that spectral methods offer, and convince young readers to courageously try out some of them. Another purpose was to make the reader aware of the importance of understanding the numerical errors involved in each algorithm, including numerous cases of spectral methods developed here.

This work can be used in undergraduate or graduate courses as a good complement to longer books such as Shizgal [1] when one analyzes physical problems in quantum mechanics, condensed matter or quantum chemistry. It can also serve as an alternative text for students to learn about topics that are analyzed in other relevant books in the area of spectral methods and their applications [2–8]. We hope to have achieved our aims.

References

1. B.D. Shizgal, *Spectral Methods in Chemistry and Physics. Applications to Kinetic Theory and Quantum Mechanics* (Springer, Dordrecht, 2015)
2. L.N. Trefethen, *Spectral Methods in MATLAB* (SIAM, Philadelphia, 2000)
3. D. Gottlieb, S.A. Orszag, *Numerical Analysis of Spectral Methods* (SIAM, Philadelphia, 1977)
4. B. Fornberg, *A Practical Guide to Pseudospectral Methods*. Cambridge Monographs on Applied and Computational Mathematics (Cambridge University Press, Cambridge, 1998)
5. J.P. Boyd, *Chebyshev and Fourier Spectral Methods* (Dover, New York, 2001)
6. G. Ben-Yu, *Spectral Methods and Their Applications* (World Scientific, Singapore, 1998)
7. D.A. Kopriva, *Implementing Spectral Methods for Partial Differential Equations Algorithms for Scientists and Engineers* (Springer, Berlin, 2009)
8. R.J. Randall, *Finite Difference Methods for Ordinary and Partial Differential Equations, Steady State and Time Dependent Problems* (SIAM, Philadelphia, 2007)

Appendix A
The MATLAB Codes for 'wave'

The program "wave" uses only "C_CM1", "SL_SR", and "mapxtor".

The subroutine "meshiem" is required for interpolation; it uses "vfg", in addition to the programs above.

Subroutine "vfg" uses for the potential a subroutine, for example "morse.m".

The function "YZ" is used when the radial interval is divided into partitions, but is not used here. The Chebyshev support points used in all our programs are defined by Eq. (3.14). They do not include the end points ± 1, contrary to their use by other authors.

Program "wave.m"

```
% start
% MATLAB example for the solution of the radial Schr. eq.
%for an exponential potential
clear all
tic
N=30; V0=-5; k=0.5; alpha = 1;
Np=1:N+1;
b1=0; b2=12; % the radial region extends from b1 to b2
[C,CM1,xz] = C_CM1(N); [SL,SR]=SL_SR(N); % are the basic matrices
r=mapxtor(b1,b2,xz); % maps the support points in x to r. Both column vectors
V = V0*exp(-r/alpha); % is the potential column vector at the support points
F = sin(k*r); G=cos(k*r);% also column vectors
plot(r,V,'*',r,F,'o',r,G,'^')
xlabel('r'), legend('V','F','G')
%pause % this is to check that all these functions are o.k.
% now we begin to solve the Lippmann-Schwinger integr. Eq.
on= ones(N+1,1); onD=mdiag(on); % is a (N+1)x(N+1)diagonal matrix=
eye(N+1)
% with 1 in the diagonal
```

© Springer Nature Switzerland AG 2018
G. Rawitscher et al., *An Introductory Guide to Computational Methods for the Solution of Physics Problems*,
https://doi.org/10.1007/978-3-319-42703-4

FD=mdiag(F); GD=mdiag(G); VD=mdiag(V); % supposed to be (N+1)x(N+1) diagonal matrices

aF=CM1*F; % is the column vector of the Cheb expansion coeffs of F

% now build up the matrices

M1 = FD*C*SR*CM1*GD + GD*C*SL*CM1*FD;

M2 = -CM1*(1/k)*((b2-b1)/2)*M1*VD*C;

M3 = onD-M2;

aPSI = M3 \ aF; % is the column of the exp coeffs of the solution PSI

semilogy(Np,abs(aF),'-o',Np,abs(aPSI),'-^')

xlabel('index');ylabel('coefficients')

legend('F','\ Psi')

pause

PSI = C*aPSI; % is the column vector of the solution at the support points

% now calculate T, and the phase shift and the normalization factor K

auxI= F.*V.*PSI;% that is the integrand of the integral involving T

aI= CM1*auxI;% are the expansion coefficients of the integrand

bI= SL*aI;% are the expansion coefficients of the indefinite integral

T = (b2-b1)*sum(bI)/(2*k);

tanphi=T, K = sqrt(T*T +1),

toc % measures the computing time from start to finish

plot(r,PSI,'-*',r,F,'-o')

xlabel('r'), ylabel('\ Psi'), legend('\Psi','F')

% it is possible to obtain PSI at regularly spaced points by interpolation

End of program wave.

Program "wave2.m"

Start of program wave2.m

% MATLAB example for the solution of the radial Schr. eq.

%for an exponential potential

clear all

tic

N=30; V0=-5; k=0.5; alpha = 1;

b1=0;

b2=20; % the radial region extends from b1 to b2

[C,CM1,xz] = C_CM1(N); [SL,SR]=SL_SR(N); % are the basic matrices

r=mapxtor(b1,b2,xz); % maps the support points in x to r. Both column vectors

V = V0*exp(-r/alpha); % is the potential column vector at the support points

F = sin(k*r); G=cos(k*r);% also column vectors

plot(r,V,'*',r,F,'o',r,G,'^')

xlabel('r'), legend('V','F','G')

%pause % this is to check that all these functions are o.k.

% now we begin to solve the Lippmann-Schwinger Integr. Eq.

on= ones(N+1,1); onD=mdiag(on); % is a (N+1)x(N+1)diagonal %matrix= eye(N+1)

```
% with 1 in the diagonal
FD=mdiag(F); GD=mdiag(G);VDD=mdiag(V); % supposed to be (N+1)x(N+1)
diagonal matrices
aF=CM1*F; % is the column vector of the Cheb expansion coeffs. of %F
CSRCM1=C*SR*CM1;CSLCM1=C*SL*CM1;
zr=zeros(N+1,1);
np = 1:N+1;
for nn=1:2
sig=(-1)^nn;
VD=sig*VDD;
% now build up the matrices
M1 = FD*CSRCM1*GD + GD*CSLCM1*FD;
M2 = -CM1*(1/k)*((b2-b1)/2)*M1*VD*C;
M3 = onD-M2;
cond_denominator=cond(M3)
cond_M2=cond(M2)
aPSI(:,nn) = M3\aF; % is the column of the exp coeffs of the solution %PSI
PSI(:,nn) = C*aPSI(:,nn); % is the column vector of the solution at %the support
points
end
toc
plot(r,PSI(:,1),'-*',r,PSI(:,2),'-o',r,zr)
xlabel('r'), ylabel('wave function'), legend('repulsive','attractive')
pause
semilogy(np,abs(aPSI(:,1)),'-*',np,abs(aPSI(:,2)),'-o')
xlabel('Chebyshev index')
ylabel('|Expansion coefficient|')
legend('repulsive','attractive')
pause
% it is possible to obtain PSI at regularly spaced points by interpolation
%toc % measures the computing time from start to finish
% end of program wave2
```

Function "C_CM1.m"

```
% start
function [C,CM1,xz]=C_CM1(N)
%f(x)=sum(1:NP1)a(n) T(j=N-1,x), j is the max power x in T(j,x)
% here there is no a_0, but a(1)= a_0/2
%calculate the coefficients a(j); j=0 to N from the Clenshaw Curtis alg.
% column vector a = CM1 * (column vector of the f's)
% column vector f at Ch. zeros of T(j=N+1) = C * (column vector a)
% xz is the column vector of the zeros of T(j=N+1)
%format short g
%format short e
```

```
% N is the highest order Chebyshev included in the expansion of f(x)
NP1=N+1;
format long e
%calc the zeros of Chebyshev of order NP1. Here m=j+1
for m=1:NP1;
theta(m)=(-1+2*m)*pi/(2*NP1);
y(m)=cos(theta(m));
%f(m)=exp(y(m));
end
xz=y';
%disp( 'the values of x and f(x)')
%[y',f'];
%Now calculate the matrix CM1(NP1,NP1) (i.e, the inverse of C)
%this requires the matrix of the T(i,j)=T(0, x0,x1,..xN);T(1,x0,x1,..xN);
for i=1:NP1;
for j=1:NP1;
T(i,j)=(cos((i-1)*theta(j)));
if i==1
A(i,j)=1/2;
else A(i,j)=T(i,j);
end
end
end
C=T';
CM1=A*(2/NP1);
%O=CM1*C;
%O;
%test if CM1 is the inverse of C.
% yes, O= unity matrix
return
end of program C_CM1
```

Function "meshiem.m"

```
%start
function [b2,err]=meshiem(N,nloop,b1,tol,rcut,k,K,xz,C,CM1)
% calculates the value of the b2, the right limit of the partition that
% starts with b1, based on the magnitudes of the last few coeffs of
% the Ch. exp of v.*f. If they are larger than "tol" the length of the
% partition is cut in half
[v,f,g] =vfg(b1,rcut,k);
%display('v(b1)')
%v
v=v/K;
kloc=sqrt(abs(k*k-v));
```

```
    if kloc < k
    kloc=k;
    end
    delta=10/kloc;
    for i=1:nloop
    del(i)=delta;
    b2t = b1+delta;
    %b2t is the trial partition right hand limit
    r = mapxtor(b1,b2t,xz);% maps the vector xz into the radial distance vector r
    [v,f,g]=vfg(r,rcut,k);
    v=v/K; %now v stands for VBAR
    av=CM1*(v);
    af=CM1*(f);
    afv=CM1*(f.*v);
    [afv];
    aux=afv(N+1)^2 + afv(N)^2 + afv(N-1)^2;
    err(i)=sqrt(aux);
    b2=b2t;
    if err(i) > tol
    delta=delta/2;
    else
    break
    end
    end
    [del',err'];
    return
% end of program "meshiem"
```

Function "mapxtor"

```
    % start
    function r=mapxtor(b1,b2,x)
    % x is the column vector of NP1 numbers in (-1,+1)
    % r is the corresponding column vector stretching from (b1, b2)
    auxm = (b2-b1)/2;
    auxp = (b2+b1)/2;
    auxr = auxp/auxm;
    r = auxm*(x+auxr);
    %display('in mapxtor')
    %auxm,auxp,x,auxr
    [auxm,auxp];
    return
    % end mapxtor
```

Function "morse.m"

```
% start
function v=morse(rr);
VBAR = 6;
alpha = 1/0.3 ;
x = (rr-4)/alpha;
y=exp(-x);
z= (y-2);
v=-VBAR*(y.*z);
return;
% end morse.m
```

Function "SL_SR"

```
% start
function [SL,SR]=SL_SR(N)
% N is the highest order Chebyshev included in the expansion of f(x)
%b=SL*a, b is the column vector of the coefficients of the expansion
%into Chb's of integral from -1 to x of f(x) dx
%c=SR*a, c is the column vector of the coefficients of the expansion
%into Chb's of integral from x to 1 of f(x) dx
%first construct the integr routines SL=M11*U11
%display(' in SL_SR')
NP1=N+1;
for i=1:NP1;
for j=1:NP1;
M11(i,j)=0;
U11(i,j)=0;
B(i,j)=0;
if i==j
M11(i,j)=1;
M22(i,j)=-1;
end
end
end
% now fill the top row
i=1; M11(i,2)=1; M22(1,1)=1; M22(1,2)=1;
for j= 3:NP1
M11(i,j)= M11(i,j-1)*(-1);
M22(i,j)= 1;
end
%disp (' M')
%M11;
%M22;
```

```
%now construct U11
aux=0;
for i=2:NP1-1
aux=aux+2;
U11(i,i+1)=-1/aux;
U11(i,i-1)=1/aux;
end
U11(2,1)=1;
U11(NP1,NP1-1)=1/(aux+2);
%disp('U; SL=M*U ')
%U11;
SL=M11*U11;
SR=M22*U11;
% These are the NP1*NP1 integration matrices
return
% end of SL_SR
```

Function "vfg"

```
% start
function [v,f,g]=vfg(r,RCUT,k)
% given the range of points in r from b1 to b2,
% then calculate the corresponding potential values
[NP1,p]=size(r);
for i=1:NP1
if(r(i)<=RCUT)
rr(i)=RCUT;
else
rr(i)=r(i);
end
end
vv=morse(rr);
v=vv';%Dv=mdiag(v);
f=sin(k*r);%Df=mdiag(f);
g=cos(k*r);%Dg=mdiag(g);
return
% end of vfgr
```

Function "YZ"

```
% start
function ... [aY,aZ,afv,OVERLAPS,errYZ]=YZ(N,b1,b2,rcut,k,K,xz,SLCM1,SR
CM1,C,CM1)
% solves the integral eq. for the Chebyshef expansion coefficients of the
% local functions Y and Z in the interval b1 <r < b2,
% N is the order of the last Chebyshev used in the expansion of Y and Z
```

```
% k = the wave number for the Green's functions
% K = h^2/(2M) in MeV fm^2.
% The potentials called by the subroutines are in MeV,
% hence VBAR=V/K and EBAR = k^2 = E/K are in fm^(-2)
%
NP1=N+1;
FACTOR=(b2-b1)/(2*k);
% (b2-b1)/2 converts the integration over dr into dx
% dividing by K converts the potential calculated in MeV by the function
% MalfTj to fm(-2) , and k is the wave number in fm^(-1). Hence, after
% the multiplication by the FACTOR, the integrals over Green's, f and pot's
% have no dimension.
I=eye(NP1);% is the unit matrix NP1xNP1
% xz is the column vector of the zeros of the Ch pol of order NP1=N+1
% af = CM1(f(xz)) is the column vector of the Ch. Coeffs of the expansion
% of f. Here af(1) is the coeff of T(0,x) , i.e., =af0/2
% f(xz)=C(af) is the column vector of the values of f(x) at the zeros xz
format short e
%[SL,SR]=SL_SR(N);
% b = SL(a).
% b is the column vector of the Ch. coeffs
% of the integral from -1 to x of f(x) dx,
% and a b is the column vector of the Ch. coeffs of the expansion of f.
% ditto for SR, for the integr from x to +1
r = mapxtor(b1,b2,xz);% maps the vector xz into the radial distance vector r
[v,f,g]=vfg(r,rcut,k);
v=v/K; %now v stands for VBAR
Dv=mdiag(v); %Dv is the diagonal matrix that has the vector v at its diagonal
Df=mdiag(f);
Dg=mdiag(g);
%display(' x r v f g');
[xz,r,v,f,g];
av=CM1*(v);
af=CM1*(f);
afv=CM1*(f.*v);
ag=CM1*(g);
%SLCM1=SL*CM1;
%SRCM1=SR*CM1;
MG = CM1*(Df*C*SRCM1*Dg + Dg*C*SLCM1*Df)*Dv*C;
M=(I+FACTOR*MG);
aY = M\af;
aZ = M\ag;
aux=aY(N+1)^2 + aY(N)^2 + aY(N-1)^2;
aux=aux + aZ(N+1)^2 + aZ(N)^2 + aZ(N-1)^2;
errYZ=sqrt(aux);
```

```
%display(' Cheb exp coeffs for v v*f Y Z f g');
[av, afv,aY,aZ,af, ag];
Y=C*aY;
Z=C*aZ;
%Now calculate the overlap integrals
% first calculate the integrals from -1 to x of g*v*z
bgvz=SLCM1*(g.*v.*Z);
bfvz=SLCM1*(f.*v.*Z);
bgvy=SLCM1*(g.*v.*Y);
bfvy=SLCM1*(f.*v.*Y);
% next sum the coefficients of the Cheb expansion of the integral, since when
% x = 1, all chebyshevs are equal to 1, and get the integral from b1 to b2.
GVZ=FACTOR*sum(bgvz);
FVZ=FACTOR*sum(bfvz);
GVY=FACTOR*sum(bgvy);
FVY=FACTOR*sum(bfvy);
OVERLAPS=[FVY;GVY;FVZ;GVZ];
return
% end of YZ
```

Appendix B
MATLAB Codes for the Derivative Matrix

Test program #2 for Eq. (11.37) in Chap. 11.

```
% test deriv2 start program 2
clear all
N=30, b1=0, b2=10
% construct sample functions fa, its first and second derivatives fpa, and
% fppa
a = 0.5; R=3.5;
%a = 0.3;
ra = 0:0.1:10;
ua = (ra-R)/a;
e1a = exp(ua); e2a = (1+e1a);
fa = (1/a)*e1a.*(e2a.^(-2));
f1a = (1/(a*a))*e1a.*e1a.*(e2a.^(-3));
fpa = fa/a - 2*f1a;
plot(ra,fa,'-*',ra,fpa,'-o')
legend ('fa','df/dr')
pause
%start the spectral method
[C,CM1,xz] = C_CM1(N); r = mapxtor(b1,b2,xz);
r=mapxtor(b1,b2,xz);
[CH,CHder1,CHder2] = der2CHEB(xz);
factor = (b2-b1)/2;
on = ones(N+1,1);
uS = (r-R*on)/a;
e1S = exp(uS); e2S = (on+e1S);
fS = (1/a)*e1S.*(e2S.^(-2));
f1S = (1/(a*a))*e1S.*e1S.*(e2S.^(-3));
fpS = fS/a - 2*f1S;
plot(ra,fa,'-*',ra,fpa,'-o')
legend ('fa','df/dr')
```

© Springer Nature Switzerland AG 2018
G. Rawitscher et al., *An Introductory Guide to Computational Methods for the Solution of Physics Problems*,
https://doi.org/10.1007/978-3-319-42703-4

```
ylabel('spectral analytic functions')
pause
af = CM1*fS; % these are the expansion coefficients for the test function fS
plot(ra,fa,r,fS,'o',ra,fpa,r,fpS,'^')
xlabel('radial distance r')
ylabel('f and df/dr')
legend ('analytic','values at Cheb mesh')
%axis([0 30 0 1])
pause
% start testing CHE
fpC= (1/factor)*CHder1*af;
% the factor 1/a is needed to transform d/du into d/dr, since u = (r-R)/a
%fppC= (1/(factor*factor))*CHder2*af;
plot(ra,fpa,'-',r,fpC,'o')
xlabel('r'); ylabel('df/dr')
legend ('anal','spectral ')
pause
% now test the accuracy
errdfdr = abs(fpS - fpC);
%errf2dr2= abs(fppS-fppC);
semilogy(r,errdfdr,'*')
xlabel('r'); ylabel('errors')
legend('df/dr')
% end program 2
```

Function der2CHEB

```
% Start function program
function [CH,CHder1,CHder2] = der2CHEB(x)
% obtains the derivative withrespect to x of Chebyshevs T(j,x)
% for j=0 to j=N at the zeros of T(N+1,x)
%result is a [1xNP1] line vector
[NP1,b] = size(x);% input x is a column vector
theta= acos(x); %theta is a column vector
sint = sin(theta);
sint2= sint.*sint;
for n=1:NP1 % j = n-1 represents the Cheb index
CH(:,n)=cos((n-1)*theta);
CHder1(:,n) = (n-1)*sin((n-1)*theta)./sint;
aux1 = (n-2)*cos((n-1)*theta);
aux2 = sin((n-2)*theta)./sint;
aux3 = (-aux1 + aux2)*(n-1)./sint2;
CHder2(:,n) = aux3;
end
%both CH and CHder are matrixes with columns along all the x values, and
```

%lines along the order of the Ch polynomial. Example CH is given below
% T_0(x_1) T_1(x_1) T_2(x_1) ... T_N(x_1)
% T_0(x_2) T_1(x_2) T_2(x_2) ... T_N(x_2)
% ....
% ....
% T_0(x_N+1) T_1(x_N+1) T_2(x_N+1) ... T_N(x_N+1)
% so CH*af = column of f at points x_1, x_2, ... x_N.
% the a's are the exp. coeffs of the function f, in column form
% the matrix CHder1 is the same as above, with the T's replaced by dT/dx
% when calling this rourine, to return to dT/dr, divide everything by
% factor = (b2-b1)/2
% So, CHder*af /factor = df/dr
% end of function program

Function "der4CHEB"

```
function [CH,CHder1,CHder2,CHder3] = der4CHEB(x)
% obtains the derivative withrespect to x of Chebyshevs T(j,x)
% for j=0 to j=N at the zeros of T(N+1,x)
%result is a [1xNP1] line vector
[NP1,b] = size(x);% input x is a column vector
theta= acos(x); %theta is a column vector
sint = sin(theta);
sint2= sint.*sint;
sint3=sint.^3; sint4=sint.^4; sint5=sint.^5;
for n=1:NP1 % j = n-1 represents the Cheb index
CH(:,n)=cos((n-1)*theta);
CHder1(:,n) = (n-1)*sin((n-1)*theta)./sint;
aux1 = (n-2)*cos((n-1)*theta);
aux2 = sin((n-2)*theta)./sint;
aux3 = (-aux1 + aux2)*(n-1)./sint2;
CHder2(:,n) = aux3;
% obtain the 3rd derivative of T_v(x), where
v = n-1;
%aux4 = v*(v-1)./sint4; aux5 = v./(sint4.*sint);
% A = (-v)*(v-1)*(v*sint.*sin(v*theta)+2*cos(v*theta).*cos(theta))./sint4;
A=(-v)*(v-1)*((v-2)*sint.*sin(v*theta)+2*cos((v-1)*theta))./sint4;
B = -v*((v+2)*cos((v-1)*theta).*sint-3*sin(v*theta))./sint5;
CHder3(:,n) = A + B;
end
```
%both CH and CHder are matrixes with columns along allthe x values, and
%lines along the order of the Ch polynomial. Example CH is given below
% T_0(x_1) T_1(x_1) T_2(x_1) ... T_N(x_1)
% T_0(x_2) T_1(x_2) T_2(x_2) ... T_N(x_2)
% ....

```
% ....
% T_0(x_N+1) T_1(x_N+1) T_2(x_N+1) ... T_N(x_N+1)
% so CH*af = column of f at points x_1, x_2, ... x_N.
% the a's are the exp. coeffs of the function f, in column form
% the matrix CHder1 is the same as above, with the T's replaced by dT/dx
% when calling this rourine, to return to dT/dr, divide everything by
% factor = (b2-b1)/2
% So, CHder*af /factor = df/dr
%end of function der4CHEB
```

Program Deriv_u

```
% Program Deriv_u
clear all
% test the derivative of u by expanding u into Chebyshevs
b1=0, b2=10,R=3.5,a = 0.5, N=50
factor = (b2 - b1)/2;
[C,CM1,xz]=C_CM1(N);
on = ones(N+1,1);
r = mapxtor(b1,b2,xz);
y = exp((r-R*on)/a);
%plot(r,y);ylabel('y')
%pause
u = y./((on+y).^2)/a;
aux = (1-2*y./(on+y));
f = aux.*u/a;
plot(r,u,'-*',r,f,'–o')
xlabel('r'),legend('u','du/dr')
%pause
% Now construct the matrix B
% call the derivative of Chebyshevs
[CH,CHder1,CHder2] = der2CHEB(xz);
B = CHder1;
M = B*CM1;
fapprox=M*u/factor;
%plot(r,fapprox,'-*',r,f,'-o')
%ylabel('du/dr')
%xlabel('r')
%legend('approx','exact')
%axis([b1 b2 -0.5 0.5])
%pause
error= abs(f-fapprox);
semilogy(r,error,'*')
xlabel('r');ylabel('error')
%axis([b1 b2 1e-7 1e1])

% test deriv via Fourier
```

```
clear all
load('Ovlp')
%[Ovlp]=Ov_cos_T(300,100);
NF = 100, NT = 60,
nfp=zeros(NF+1,1);,
nfp2=zeros(NF+1,1);
for n = 1:NF+1
nfp(n)=n-1 ;
nfp2(n)=(n-1)^2;
end
for n=1:NT+1
nct(n) =n-1;
end
M=Ovlp(1:NF+1,1:NT+1);
[C,CM1,xz]=C_CM1(NT);
[SL,SR]=SL_SR(NT);
yz = acos(xz);
on = ones(NT+1,1);
% test function u(r)=(1/a)y/(1+y)
%a=0.5,b1=0,b2=10,R=4
b1=0,b2=pi, factor = (b2-b1)/2;
r=mapxtor(b1,b2,xz);
rhalf= r.^(1/2);
u = rhalf.*cos(r);
% the derivative du/dr= cos(r)./(2*rhalf)-rhalf*sin(r)
dudr=cos(r)./(2*rhalf)-rhalf.*sin(r);
% now calculate d2udr2
term1 = 0.25*(r.^(-3/2))+rhalf;
d2udr2 =-term1.*cos(r) - sin(r)./rhalf;
plot(r,u,'*',r,dudr,'-o',r,d2udr2,'-`');
xlabel('r')
legend('u','du/dr','d2u/dr2')
axis([b1 b2 -5 5])
pause
plot(xz,u,'*',xz,dudr,'-o')
ylabel('u, du/dr'), ylabel('x')
%pause
a=CM1*u;
semilogy(nct,abs(a),'*')
xlabel('n'); ylabel('|a_n(u)|')
pause
c= M*a;%factor;
c(1) = 0.5*c(1);
%plot(nfp,c)
c1 =((pi/2)/(factor))*(c.*nfp);
```

```
c2=((pi/2)/(factor))^2*(c.*nfp2);
pause
%hold on
semilogy(nfp,abs(c),'*',nfp,abs(c1),'o',nfp,abs(c2),'x')
xlabel('nf'), ylabel('Fourier exp. coeffs')
legend('c','c1','c2')
pause
% now calculate the first derivative of u at the Cheb support points
[phic,xz,r]=PHICOS(b1,b2,NF,NT);
uf= phic(:,:)*c;
plot (r,uf,'o',r,u,'-')
legend('fourier','Chebyshev')
xlabel('r'), ylabel('u')
pause
erroru=abs(uf-u);
semilogy(r,erroru,'*')
ylabel('|uf-u|'), xlabel('r')
pause
[phis,xz,r]=PHISIN(b1,b2,NF,NT);
duf= (-1)*phis(:,:)*c1;
plot (r,duf,'o',r,dudr,'-.')
xlabel('r'), ylabel('du/dr')
legend('fourier','Chebyshev')
%pause
errordu=abs(duf-dudr);
semilogy(r,errordu,'*')
ylabel('|du/drf- du/dr|'), xlabel('r')
pause
d2uf=(-1)*phic(:,:)*c2;
plot (r,d2uf,'.',r,d2udr2,'-o')
legend('fourier','Chebyshev')
xlabel('r'), ylabel('d2u/dr2')
pause
errord2u= abs(d2uf-d2udr2);
semilogy(r,errord2u,'-*')
xlabel('r'), ylabel('d2u/dr2 (four-anal)')
pause
% now calc the deriv from the deriv of the Chebs
[CH,CHder1,CHder2] = der2CHEB(xz);
f = CHder2;
%[d3,d4,d5] = der3CHEB(b1,b2,f,NG );
%plot(r,CH(:,4));
    plot(r,CHder1(:,6),'*',r,CHder2(:,6),'o'); % that is the second deriv of Chebs,
column T4(r=b1,r2, ..rN)
    xlabel('r'), ylabel ('CHder')
```

```
legend('der1','der2')
% calc the derv a la CHEB
d1uc=CHder1*a/factor;
d2uc=CHder2*a/(factor^2);
plot (r,d1uc,'*',r,d2uc,'o')
xlabel('r'),ylabel('Cheb der of u')
legend('der 1','der 2')
pause
errCHEBdu=abs(d1uc-dudr);
errCHEBd2u= abs(d2uc-d2udr2);
semilogy(r,errCHEBdu,'*',r,errCHEBd2u,'o',r,errord2u,'^')
xlabel('r'), ylabel('errors')
legend('CH-1','CH-2','FO-2')
pause
%axis([ b1 b2 1e-2 10])
% test v = r^(1/2)sin(r)
v = rhalf.*sin(r);
cv = CM1*v;
plot (r,v)
semilogy(nct,abs(cv))
```

Program test_f1f2

```
% test deriv via Fourier and hybrid
clear all
load('Ovlp')
%[Ovlp]=Ov_cos_T(300,100);
NF = 100, NT = 60,% was 100 and 60
nfp=zeros(NF+1,1);,
nfp2=zeros(NF+1,1);
for n = 1:NF+1
nfp(n)=n-1 ;
nfp2(n)=(n-1)^2;
end
for n=1:NT+1
nct(n) =n-1;
end
M=Ovlp(1:NF+1,1:NT+1);
[C,CM1,xz]=C_CM1(NT);
[SL,SR]=SL_SR(NT);
yz = acos(xz);
on = ones(NT+1,1);
% test function u(r)=(1/a)y/(1+y)
%a=0.5,b1=0,b2=10,R=4
```

```
b1=3,b2=10, factor = (b2-b1)/2;
b1=0,b2=pi, factor = (b2-b1)/2;
r=mapxtor(b1,b2,xz);
rhalf= r.^(1/2);
u = rhalf.*cos(r);
% the derivative du/dr= cos(r)./(2*rhalf)-rhalf*sin(r)
dudr=cos(r)./(2*rhalf)-rhalf.*sin(r);
% now calculate d2udr2
term1 = 0.25*(r.^(-3/2))+rhalf;
d2udr2 =-term1.*cos(r) - sin(r)./rhalf;
% now calculate d3u/dr3
auxc = (3/8)*r.^((-5/2))-(3/2)*(r.^(-1/2));
auxs = (3/4)*(r.^(-3/2)) + r.^(1/2);
d3udr3= cos(r).*auxc +sin(r).*auxs;
plot(r,u,'*',r,dudr,'-o',r,d2udr2,'-^',r,d3udr3,'-d');
xlabel('r'),% ylabel('analytic')
legend('v','dv/dr','d^2v/dr^2','d^3v/dr^3')
axis([b1 b2 -5 5])
pause
plot(xz,u,'*',xz,dudr,'-o')
ylabel('u, du/dr'), xlabel('x')
pause
a=CM1*u;
semilogy(nct,abs(a),'*')
xlabel('n'); ylabel('la_n(v)l')
pause
uCH= C*a;
errCH= abs(u - uCH);
semilogy(r,errCH,'*')
xlabel('r')
ylabel('u-uCH')
pause
c= M*a;%factor;
c(1) = 0.5*c(1);
%plot(nfp,c)
c1 =((pi/2)/(factor))*(c.*nfp);
c2=((pi/2)/(factor))^2*(c.*nfp2);
pause
%hold on
semilogy(nfp,abs(c),'*',nfp,abs(c1),'o',nfp,abs(c2),'x')
xlabel('nf'), ylabel('Fourier exp. coeffs')
legend('c','c1','c2')
pause
% now calculate the first derivative of u at the Cheb support points
[phic,xz,r]=PHICOS(b1,b2,NF,NT);
```

```
uf= phic(:,:)*c;
plot (r,uf,'o',r,u,'-')
legend('fourier','Chebyshev')
xlabel('r'), ylabel('u')
pause
erroru=abs(uf-u);
semilogy(r,erroru,'*')
ylabel('|uf-u|'), xlabel('r')
pause
[phis,xz,r]=PHISIN(b1,b2,NF,NT);
duf= (-1)*phis(:,:)*c1;
plot (r,duf,'o',r,dudr,'-.')
xlabel('r'), ylabel('du/dr')
legend('fourier','Chebyshev')
%pause
errorduf=abs(duf-dudr);
semilogy(r,errorduf,'*')
ylabel('|du/drf- du/dr|'), xlabel('r')
pause
d2uf=(-1)*phic(:,:)*c2;
plot (r,d2uf,'.',r,d2udr2,'-o')
legend('fourier','Chebyshev')
xlabel('r'), ylabel('d2u/dr2')
pause
errord2uf= abs(d2uf-d2udr2);
semilogy(r,errord2uf,'-*')
xlabel('r'), ylabel('d2u/dr2 (four-anal)')
pause
% now calc the deriv from the deriv of the Chebs
%[CH,CHder1,CHder2] = der2CHEB(xz);
[CH,CHder1,CHder2,CHder3] = der4CHEB(xz);
f = CHder2;
%[d3,d4,d5] = der3CHEB(b1,b2,f,NG );
%plot(r,CH(:,4));
plot(r,CHder1(:,6),'*',r,CHder2(:,6),'o'); % that is the second deriv of Chebs,
column T4(r=b1,r2, ..rN)
xlabel('r'), ylabel ('CHder')
legend('der1','der2')
% calc the derv a la CHEB ***
d1uc=CHder1*a/factor;
d2uc=CHder2*a/(factor^2);
d3uc=CHder3*a/(factor^3);
ac= CM1*d1uc; %prepare to do the second derivative
a2c=CM1*d2uc; % prepare for the thrd deruivative
a3c=CM1*d3uc; % prepare for the thrd deruivative
```

```
d2ucc = CHder1*ac/factor; % is the second derivative calc from the firs der.
%d3uc = CHder1*a2c/factor; % is the third derivative calc from the 2nd der.
a3c=CM1*d3uc;
semilogy(nct,abs(ac),'^',nct,abs(a2c),'o',nct,abs(a3c),'*')
xlabel('n'), ylabel('derivative ch exp coeffs')
legend('du/dr','d2u/dr2','d3u/dr3')
axis([1 NT 1e-3 1e6])
pause
plot (r,d1uc,'*',r,d2uc,'o',r,d3uc,'-.')
xlabel('r'),ylabel('der of u via Cheb')
legend('der 1','der 2','der 3')
% the second deriv calc from the first, CHder1 * CHder1
% gives identical errors to using CHEder2
axis([b1 b2 -5 5])
pause
errCHEBdu=abs(d1uc-dudr);
errCHEBd2u= abs(d2uc-d2udr2);
errCHEBd3u= abs(d3uc-d3udr3);
semilogy(r,errCHEBdu,'*',r,errCHEBd2u,'o',r,errCHEBd3u,'-.',r,errord2uf,'^')
xlabel('r'), ylabel('errors')
legend('CH-1','CH-2','CH-3','FO-2')
pause
%axis([ b1 b2 1e-2 10])
% test v = r^(1/2)sin(r)
v = rhalf.*cos(r);
cv = CM1*v;
plot (r,v)
%pause
semilogy(nct,abs(cv))
% end of program test_f1f2
```

Program test_hybrid_deriv3

```
%Test_rmesh
clear all
b1 = 0; b2 = 3, N=60;bp(1) = b2; mp(1) = 1;h=1e-2;
b1 = 3; b2 = 10, N=60;bp(1) = b2; mp(1) = 1;h=1e-2;
for n = 1:N+1
ntp(n)=n;
end
[C,CM1,xz] = C_CM1(N); [SL,SR]=SL_SR(N);
bp(1) = b2; mp(1) = 1;
r=mapxtor(b1,b2,xz);
rhalf= r.^(1/2);
u = rhalf.*cos(r);
```

```
% the derivative du/dr= cos(r)./(2*rhalf)-rhalf*sin(r)
dudr=cos(r)./(2*rhalf)-rhalf.*sin(r);
% now calculate d2udr2
term1 = 0.25*(r.^(-3/2))+rhalf;
d2udr2 =-term1.*cos(r) - sin(r)./rhalf;%Test_rmesh
% now calculate d3u/dr3
auxc = (3/8)*r.^((-5/2))-(3/2)*(r.^(-1/2));
auxs = (3/4)*(r.^(-3/2)) + r.^(1/2);
d3udr3= cos(r).*auxc +sin(r).*auxs;
plot(r,u,'*',r,dudr,'-o',r,d2udr2,'-^',r,d3udr3,'-d');
xlabel('r'),% ylabel('analytic')
legend('v','dv/dr','d^2v/dr^2','d^3v/dr^3')
axis([b1 b2 -5 5])
pause
au = CM1*u;
np = 1; % there is only one partition
% Hybrid, first derivative
for n= 3:N-2
ri(1) = r(n)- h; ri(2)= r(n);ri(3)=r(n)+h;ri(4)= r(n)+2* h ;
[npr,b1,b2,x,ch]=rimesh(ri,b1,np,bp,N);
ui= ch*au;
d1udr1i(1,n)= (-2*ui(1)-3*ui(2)+6*ui(3)- ui(4))/(6*h);
end
plot(r,dudr,'o',r(3:N-2),d1udr1i(3:N-2)','.')
axis([b1 b2 -3 5])
xlabel('r'),ylabel('d3udr3')
legend('analytic','finite diff')
%pause
error1i =abs(dudr(3:N-4)- d1udr1i(3:N-4)');
semilogy(r(3:N-4),error1i,'*')
xlabel('r'),ylabel('error du/dr finite diff')
%axis([b1 b2 1e-6 1e-4])
h=1e-2;d2udr2i=zeros(1,N+1);
n=1;ri(1)=r(n);ri(2)=r(n)-h;ri(3)=r(n)-2*h;
[npr,b1,b2,x,ch]=rimesh(ri',b1,np,bp,N);
%[nrm,b1,b2,x,ch]=rimesh(r,bstart,np,bp,N)
ui1= ch*au;
%d2udr2i(1,1)=(ui1(1)-2*ui1(2)+ui1(3))/((h^2));
%[ui1']
n=N+1;ri(1)=r(n)+3*h;ri(2)=r(n)+2*h;ri(3)=r(n)+h;
[npr,b1,b2,x,ch]=rimesh(ri',b1,np,bp,N);
uiN= ch*au;
d2udr2i(1,N+1)=(uiN(1)-2*uiN(2)+uiN(3))/((h^2));
%[uiN']
% [ri'],[uiN],d2udr2i
```

```
% pause
%ri=zeros(3,1); ch=zeros(3,16);
for n= 2:N
ri(1) = r(n) - h; ri(2) = r(n) ; ri(3) = r(n) + h ;
[npr,b1,b2,x,ch]=rimesh(ri,b1,np,bp,N);
ui= ch*au;
d2udr2i(1,n)= (ui(1)-2*ui(2) +ui(3))/h^2;
end
d2udr2i(1,N+1)=d2udr2i(1,N);
d2udr2i(1,1)=d2udr2i(1,2);
plot(r,d2udr2,'o',r,d2udr2i,'.')
axis([b1 b2 -3 5])
xlabel('r'),ylabel('d2udr2')
legend('analytic','finite diff')
pause
%h = ((24*(10^-6))^(1/4))
errori2 =abs(d2udr2- d2udr2i');
errori2(1,1)=errori2(2,1);errori2(61,1)=errori2(60,1);
semilogy(r,errori2,'o')
ylabel('error du2/dr2 finite diff')
pause
expecterror = (h^3)/2
%semilogy(r(2:N),error(2:N),'*')
semilogy(r,errori2,'*')
xlabel('r'), ylabel('|d2udr2i-d2urdr2 anal| ')
ad2ui=CM1*d2udr2i';
semilogy(ntp, abs(ad2ui),'^',ntp,abs(au),'o')
legend('ad2i','au')
% now calculate the third derivative via the hybrid method
for n= 3:N-2
ri(1) = r(n)- 2*h; ri(2)= r(n)-h;ri(3)=r(n);ri(4)= r(n)+ h ; ri(5)=r(n)+2*h;
[npr,b1,b2,x,ch]=rimesh(ri,b1,np,bp,N);
ui= ch*au;
d3udr3i(1,n)= (-2*ui(1)+4*ui(2)-4*ui(4)+ 2*ui(5))/(4*h^3);
end
plot(r,d3udr3,'o',r(3:N-2),d3udr3i(3:N-2),'.')
axis([b1 b2 -3 5])
xlabel('r'),ylabel('d3udr3')
legend('analytic','finite diff')
pause
errori3 = abs(d3udr3(3:N-2)-d3udr3i(3:N-2)');
errori2 =abs(d2udr2- d2udr2i');
errori2(1,1)=errori2(2,1);errori2(61,1)=errori2(60,1);
semilogy(r(3:N-4),error1i,'*',r,errori2,'o',r(3:N-4),errori3(3:N-4),'^')
ylabel('error hybrid'),xlabel('r')
```

```
legend('dv/dr2','d^2v/dr^2','d^3v/dr^3')
%axis([b1 b2 1e-6 1e-3])
pause
%expecterror = (h^3)/2
semilogy(r(3:N-4),errori3(3:N-4),'*')
xlabel('r'),ylabel('error du3/dr3 finite diff')
%axis([b1 b2 1e-6 1e-4])
%end of program test_hybrind_deriv3
```

Program Ov_cos_T

```
% overlap < phi(nF)*T(nT > = integral -1 to +1
% (cos((1+x)(pi/2)nF)*cos(nt*theta)) dx
function [Ovlp_cos]=Ov_cos_T(NF,NT);
NS= NF + NT;
[C,CM1,xz]=C_CM1(NS);
[SL,SR]=SL_SR(NS);
on = ones(NS+1,1);
for nf=1:NF+1
cs(nf,:)=cos((-on+xz)*pi*(nf-1)/2);
end
for nt=1:NT+1
T(nt,:)=cos((nt-1)*acos(xz));
end
for nf = 1:NF+1
for nt = 1:NT+1
aux1= cs(nf,:).*T(nt,:);
a1=CM1*aux1';
b1=SL*a1;
Ovlp_cos(nf,nt)=sum(b1);
end
end
save('Ovlp')
% end of program Ov_cos_T
```

Program Ov_sin_T

```
% overlap < phi(nF)*T(nT > = integral -1 to +1
% (cos((1+x)(pi/2)nF)*cos(nt*theta)) dx
function [Ovlp_sin]=Ov_sin_T(NF,NT);
NS = NT + NF;
[C,CM1,xz]=C_CM1(NS);
[SL,SR]=SL_SR(NS);
on = ones(NS+1,1);
```

```
for nf=1:NF+1
cs(:,nf)=sin((-on+xz)*pi*(nf-1)/2);
end
for nt=1:NT+1
T(:,nt)=cos((nt-1)*acos(xz));
end
for nf = 1:NF+1
for nt = 1:NT+1
aux1= cs(:,nf).*T(:,nt);
%if(nf==3)
% plot(aux1)
%end
% pause
a1=CM1*aux1;
b1=SL*a1;
Ovlp_sin(nf,nt)=sum(b1);
end
end
save('Ovlp')
% end of program Ov_sin_T
```

Index

© Springer Nature Switzerland AG 2018
G. Rawitscher et al., *An Introductory Guide to Computational Methods
for the Solution of Physics Problems*,
https://doi.org/10.1007/978-3-319-42703-4

Printed in the United States
By Bookmasters